Building Equitable Climate Adaptation Partnerships in the US Caribbean

Sarah Molinari · Mimi Sheller
Editors

Building Equitable Climate Adaptation Partnerships in the US Caribbean

palgrave
macmillan

Editors
Sarah Molinari
Worcester Polytechnic Institute
Worcester, MA, USA

Mimi Sheller
Worcester Polytechnic Institute
Worcester, MA, USA

ISBN 978-3-032-15726-3 ISBN 978-3-032-15727-0 (eBook)
https://doi.org/10.1007/978-3-032-15727-0

This Palgrave Macmillan imprint is published by the registered company Springer Nature Switzerland AG
The registered company address is: Gewerbestrasse 11, 6330 Cham, Switzerland

Acknowledgments We want to thank all of our collaborators in CCAN who have supported this work and helped make it possible. Led by Principal Investigator Pablo Méndez-Lázaro, the full list of members can be found at http://ccan-upr.org. We thank the many community organizations who have partnered with the WPI Puerto Rico Project Center (https://wp.wpi.edu/puertorico/) during the course of this research. We thank CCAN co-principal investigators Kim Waddell and Greg Guannel at University of the Virgin Islands and our other dedicated collaborators in the US Virgin Islands. We also thank our colleagues in the Global School at WPI, including the dedicated faculty and staff involved in the Global Projects Program, and operations management and administrative assistance from Dawn Farmer and Nancy Fay.

Competing Interests This study was partially supported through the Caribbean Collaborative Action Network (CCAN), a NOAA Climate Adaptation Partnerships team. Grant Number NA22OAR4310545 (PI, Pablo Méndez-Lazaro). It was also supported in part by the NOAA BIL Grant Number NA23OAR4310396-T1-01, Improving Engagement Methods for Coastal Resilience and Reducing Climate Risk: Bridging Learning Networks from the Urban Northeast (CCRUN) to the US Caribbean (CCAN) (PI, Mimi Sheller, WPI).

Ethics Approval Approval for this research was granted by the University of Puerto Rico, Medical Sciences Campus, Human Research Subjects Protection Office, Institutional Review Board.

(Date Renewal 9/11/2024, No. 2304097058R001).

Contents

Notes on Contributors

Jonathan Chang is a graduate of the Community Climate Adaptation master's program at Worcester Polytechnic Institute and a former Fulbright Scholar. Originally from Samoa, he is an environmental scientist at the Scientific Research Organization of Samoa (SROS), where his work focuses on climate resilience, greenhouse gas reporting and environmental policy analysis. He is committed to supporting island communities through research, capacity building and international collaboration.

John-Michael Davis, Ph.D., is an Assistant Professor of Teaching in the Department of Integrated and Global Studies at Worcester Polytechnic Institute (WPI). His research centers on community-driven, action-oriented approaches to community resilience and sustainable development. From 2012 to 2019, he conducted a participatory action research project within the Israeli-Palestinian e-waste system, generating insights that informed policy and catalyzed systemic change. Since 2020, he has co-directed WPI's Puerto Rico Project Center, facilitating transformative student experiences through Shared Action Learning projects. His work emphasizes long-term partnerships, mutual learning, and tangible contributions that support grassroots organizations in advancing social and environmental justice.

Sol Giesso is a graduate of the Community Climate Adaptation master's program at Worcester Polytechnic Institute. She is passionate about the intersections of climate change, migration, and education and has been awarded for her contributions at the intersections of research and practice. She is pursuing an Ed.M. in Teaching and Teacher Leadership at the Harvard Graduate School of Education with the goal of furthering environmental education and civic engagement in youth.

Scott Jiusto, Ph.D., is Professor Emeritus in the Department of Integrated and Global Studies at Worcester Polytechnic Institute (WPI). His work integrates teaching, research, and community engagement through project-based learning. From 2007 to 2022, he led student teams in Cape Town, Puerto Rico, and other global sites, advancing sustainable development through Shared Action Learning. His research explores informal settlement upgrading, water and sanitation, and early childhood development, with a focus on collaborative, reflective practice. As co-director of WPI's Puerto Rico Project Center, he helped shape long-term partnerships that support grassroots organizations in building social infrastructure and advancing climate justice.

Bárbara López-González, Ph.D., is a geographer whose work engages with the political ecology of disasters, community *autogestión* (self-management), and the pursuit of transformations and just and sustainable futures. She holds a Ph.D in Geography and International Development from the University of Guelph in Canada. Bárbara currently serves as a Postdoctoral Fellow at Worcester Polytechnic Institute in Massachusetts, where she is conducting research with the Caribbean Collaborative Action Network (CCAN). Over the last six years, her work has been grounded in Caribbean and Latin American contexts, employing qualitative, ethnographic, and participatory methodologies, as well as community-engaged research and service-learning approaches.

Sarah Molinari, Ph.D., is an Assistant Teaching Professor in the Department of Integrative and Global Studies at Worcester Polytechnic Institute and a Faculty Researcher with the Caribbean Collaborative Action Network. She holds a Ph.D. in Anthropology from the CUNY Graduate Center and works at the intersections of ethnography, critical environmental studies, feminist geography, and digital humanities. Sarah's research broadly examines the politics and lived experiences of debt, mutual aid, and the climate crisis in Puerto Rico and the greater Caribbean region. She is also a co-founder of the Puerto Rico Syllabus.

Mimi Sheller, Ph.D., is Dean of the Global School at Worcester Polytechnic Institute, in Massachusetts. Sheller is an interdisciplinary social scientist with work in Caribbean Studies, Mobilities Research, and Social Theory. Sheller was founding co-editor of the journal *Mobilities*, and past President of the International Association for the History of Transport, Traffic and Mobility. She has published more than 150 articles and chapters, and recent books include *Advanced Introduction to Mobilities* (Edward Elgar, 2021); *Island Futures: Caribbean Survival in the Anthropocene* (Duke University Press, 2020); and *Mobility Justice: The Politics of Movement in an Age of Extremes* (Verso, 2018).

Sarah Strauss, Ph.D., is Professor of Anthropology in the Department of Integrative and Global Studies at Worcester Polytechnic Institute (WPI) in Massachusetts, where she was the inaugural director of the graduate program in Community Climate Adaptation. Strauss has degrees in comparative religion, public health, and cultural anthropology; her research focus lies at the intersection of environmental and health values and practices, especially those related to water, forests, climate change, and energy transitions. Her books include *Positioning Yoga* (2005), *Weather, Climate, Culture* (2004, edited with Ben Orlove) and *Cultures of Energy* (2013, edited with Stephanie Rupp and Thomas Love).

Seth P. Tuler, Ph.D., is an Associate Professor in the Department of Integrative and Global Studies Division, Worcester Polytechnic Institute, MA. Seth's research interests focus on risk governance, public participation in risk assessment and decision making, and developing tools to characterize human impacts and vulnerabilities to risk events. He has extensive experience with interdisciplinary research in multiple policy arenas, including climate adaptation planning, energy-related infrastructure siting, nuclear waste management, and regional land-use planning. A thread throughout his work has been to empower communities to participate more effectively in the management of risk.

Abbreviations

AAFAF	Fiscal Agency and Financial Advisory Authority
CAM-LC	Centro de Apoyo Mutuo Las Carolinas
CAP	Climate Adaptation Partnerships
CBO	Community-based Organization
CCA	Community Climate Adaptation
CCAN	Caribbean Collaborative Action Network
CEBDI	Evaluation and Transfer of Real Estate Assets Committee
CLWUP	Comprehensive Land and Water Use Plan
DIGS	Department of Integrative and Global Studies
DPNR	Department of Planning and Natural Resources
DTOP	Department of Transportation and Public Works
EDC	Economic Development Commission
EWB	Engineers Without Borders
FEMA	Federal Emergency Management Agency
FPR	Foundation for Puerto Rico
HMRP	Hazard Mitigation and Resilience Plan
IQP	Interactive Qualifying Project
NGO	Non-Governmental Organization
NOAA	National Oceanic and Atmospheric Administration
PR	Puerto Rico
PRPC	Puerto Rico Project Center
PV	Participatory Video
SAL	Shared Action Learning
SETS	Social-ecological-technological systems
STEM	Science, Technology, Engineering, and Math
UPR	University of Puerto Rico

USVI	United States Virgin Islands
UVI	University of the Virgin Islands
VCAPS	Vulnerability, Consequences, Adaptation Planning Scenarios
VI-EPSCoR	Virgin Islands Experimental Program to Stimulate Competitive Research
VIPA	Virgin Islands Port Authority
VISCO	Virgin Islands Studies Collective
WAPA	Water and Power Authority
WCRP	Whole Communities Resilience Planning Program
WPI	Worcester Polytechnic Institute

List of Figures

Chapter 4

Chapter 5

Chapter 6

CHAPTER 1

Introduction

Mimi Sheller and Sarah Molinari

Abstract This chapter introduces the research questions informing this book, arising out of the work of Worcester Polytechnic Institute researchers who are part of the Caribbean Collaborative Action Network, a NOAA Climate Adaptation Partnership team. After discussing the key questions framing this collection, we reflect on the power dynamics of our research in the US Caribbean context of Puerto Rico and the US Virgin Islands in relation to climate mobilities (as well as immobilities and dwelling). An overview of the chapters follows, each of which exemplifies a specific university-community-research partnership for the co-production of knowledge and climate action.

M. Sheller (✉)
The Global School, Worcester Polytechnic Institute, Worcester, MA, USA
e-mail: msheller@wpi.edu

S. Molinari
Department of Integrative and Global Studies, Worcester Polytechnic Institute, Worcester, MA, USA
e-mail: smolinari@wpi.edu

S. Molinari and M. Sheller (eds.), *Building Equitable Climate Adaptation Partnerships in the US Caribbean*,
https://doi.org/10.1007/978-3-032-15727-0_1

Keywords Climate adaptation · Co-design · Ethics · STEM education · US Caribbean

1 Introduction

This book arose out of the research of the Caribbean Collaborative Action Network (CCAN), a NOAA Climate Adaptation Partnership team representing the US Caribbean region, serving Puerto Rico and the US Virgin Islands. Led by researchers at the University of Puerto Rico and the University of the Virgin Islands, along with several other institutional partners in the continental United States, CCAN formed as a learning network that seeks to translate knowledge-to-action, and back again. The overarching goal of CCAN is to foster knowledge sharing for more sustainable and equitable strategies for co-production of climate action with communities throughout Puerto Rico and the USVI. We discuss the network and its project in Chapter 2. This book does not address the CCAN project as a whole but rather seeks to reflect on a more specific aspect: the experiences of part of the research team based at Worcester Polytechnic Institute (WPI) in Massachusetts, in order to open a discussion of the practical, methodological, and ethical considerations within such collaborative partnerships. This volume also highlights the strands of qualitative research and community-engaged learning within the CCAN network, which also includes many other science activities.

We believe that critical self-reflexivity is crucial for moving toward more equitable research partnerships that potentially might support decolonial epistemologies—without claiming that we have achieved such a goal yet. This book will focus on questions of best practices for building equitable and ethical approaches for shared action research and community-research partnerships among academic researchers, community-based organizations (CBOs), local and federal governmental agencies, and student research teams who are coming together in support of marginalized communities facing high risks and high vulnerability to climate change impacts. This work within communities made vulnerable to the impacts of climate change calls on us (as a research team based largely in New England, with a presence in the region through WPI's Puerto Rico Project Center and other family, research, and collaborative ties) to address questions of power, equity, and justice within our own research methodologies and research activities. It also demands that we consider the extent to which our work, and the wider funding environment that supports it, is furthering climate justice (and other dimensions of justice), or not.

Some academic researchers working in the field of climate justice argue that there is a need for universities to lead more transformative change toward climate justice, leveraging resources and capacities toward reparative ends. Jennie C. Stephens argues that "Amid the worsening climate crisis and intensifying inequities, higher education can play a powerful role in addressing the intersecting crises facing humanity. Institutions of higher education hold untapped potential to advance social justice and reduce climate injustices," but currently this is not happening enough (Stephens, 2024). Therefore, we situate this book within wider questions being asked of universities today, including: How can institutional structures such as existing or new degree programs, or access to federal research grants, be shaped toward these goals? Can research funding be utilized in ways that support greater climate justice? Is student project-based learning an effective way to build community partnerships for climate adaptation? How can "co-production" of climate knowledge and adaptation strategies with non-academic community partners support multiple ways of knowing and producing knowledge about climate hazards and more just futures? And how does this undertaking add to ethical Science, Technology, Engineering, and Math (STEM) education in this era of human-driven climate change?

Today (in early 2026), we find ourselves in a dynamic situation not only of climate instability, but also institutional turbulence and transformations in the governance of climate change. The Caribbean region has experienced a series of unprecedented category five hurricanes with devastating impacts on highly vulnerable communities. This is an especially difficult moment of institutional, political, and financial stress that is affecting not only frontline communities who are disproportionately experiencing the impacts of climate change and environmental hazards, but also those who might seek to help them. In the United States especially, the higher education system itself and those who study climate change are under pressure, as well as federal research funders such as the National Oceanic and Atmospheric Administration (NOAA), the source of CCAN funding. The traditional expectations that elite "R1" universities have had for the continuation of federally-funded research programs are no longer viable due to recent changes in the federal government.

Meanwhile, historically underserved institutions (like our partners at the University of Puerto Rico and the University of the Virgin Islands) may find they have even less access to funding opportunities. The impacts of limited federal research funds will be unevenly distributed across institutions and have the potential to disrupt research collaborations between better resourced universities and institutions already hit with years of austerity.

These pressures on researchers, institutions, and frontline community participants occur at the very moment that the research paradigm has been shifting toward stronger and deeper participatory action research and support for more equitable knowledge-action networks. Much current research on climate adaptation and resilience calls for not only new convergent interdisciplinary research approaches but also the co-creation with communities of effective systems to preempt, prepare for, track, and respond to climate-induced hazards and vulnerabilities. Critical participatory approaches call for meaningful engagement with stakeholders in every aspect of the research process. Hendricks et al. (2022, p. 2), for example, call for the use of community science and public participation to "empower residents in using community-driven research, analytics, organizing, and mobilization to advocate on their own behalf and to implement community-level programs that level the landscape before, during, and after flood disasters."

The challenge is how to do participatory action research in ways that pay attention to power and inequities in terms of who can speak, whose voice is heard, who exercises agency, and who can make decisions. Black feminist planning approaches further take into account structural racism and intersectional inequalities. Fayola Jacobs highlights first that "systemic forms of racism, sexism and classism severely impact communities' abilities to prepare for and recover from disaster" and that radical planning therefore requires alternative methods of inclusion that acknowledge that "lived experience is a way of knowing and knowledge is created and validated through dialogue" (Jacobs, 2019, p. 26). This offers a moment to reflect creatively on how university researchers and students might continue to work with community-based organizations and local governments, and especially with the most marginalized regions of the United States that have already been experiencing the withdrawal of government services for many years. What should be the role of universities in supporting community action in adapting to climate change and building new ways of living together at this turning point? Each of the research projects described in the following chapters is concerned with supporting different kinds of social infrastructure for creative climate adaptation, co-designing it together with community partners.

In reflecting on the WPI social science research team's contributions to CCAN and to ethical STEM education involving community partnerships, we seek to advance collective learning around how to translate community-engaged research into implementation that supports equity and justice, with real-world outcomes that build community capacities to define their own future. We also question and critically reflect on the ability of our own research approaches and methodologies to achieve this, even under the best of circumstances. We seek to evaluate both the strengths and weaknesses of participatory approaches and emergent human-centered co-design methodologies for furthering more equitable outcomes of climate research and climate governance. In the rest of this introduction, we introduce the United States Caribbean territories in the context of governance of climate change risks and we briefly summarize the chapters that follow.

2 The US Caribbean Context and Climate (Im)mobilities

Across the Caribbean region, climate change-related risks are increasing with a heightened potential for damage to people and infrastructure (McPhillips et al., 2018). Caribbean island communities such as the US Virgin Islands (USVI) and Puerto Rico (PR) (see Fig. 1) are increasingly exposed to severe weather, including tropical cyclones, heat waves, and droughts, with impacts that are compounded by limited supply chains, weak infrastructure, and governance systems burdened by layers of bureaucracy that often lead to inaction (Gonzalez et al., 2017; Makoff & Setser, 2017). During and after Hurricanes Irma and Maria in 2017, communities in both US Caribbean territories experienced major disruptions in essential services (such as water supply, sanitation, and power) and the degradation of environmental and human health, which contributed to high rates of mortality. These island communities additionally lack adequate resources with which to prepare, respond, and recover from ongoing climate crises.

The US territories of PR and the USVI have still not fully recovered from the 2017 hurricane season. The impacts of weather extremes on essential services and critical infrastructure (education, health, power,

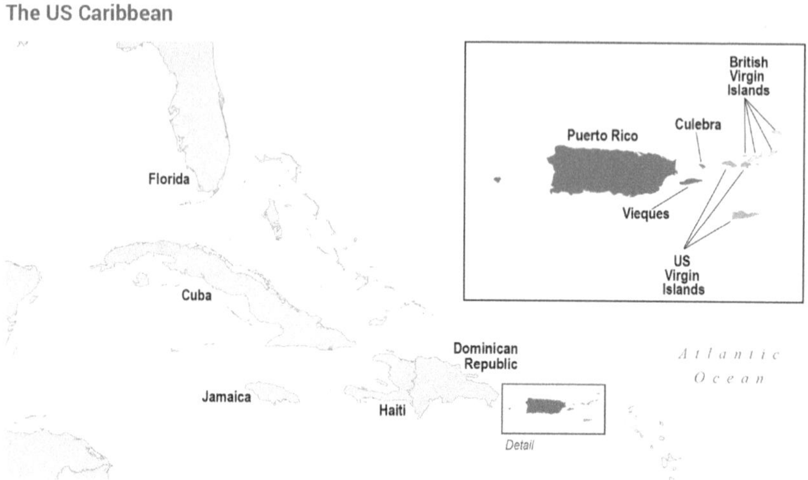

Fig. 1 Map of the US Caribbean. The map shows the location of Puerto Rico and the US Virgin Islands in the Caribbean Basin. The inset shows the main island of Puerto Rico (green; the two municipal islands of Vieques and Culebra are shown to the east and Mona Island to the west) and the US Virgin Islands (yellow; St. Thomas and St. John to the north and St. Croix to the south). Figure credit: University of Puerto Rico, NOAA NCEI, and CISESS NC. Adapted From Fifth National Climate Assessment, Chapter 23, US Caribbean (Méndez-Lazaro et al., 2023, pp. 23–6)

water, etc.) have led to significant population losses [−18% in the USVI; −12% in PR since 2010 (U.S. Census Bureau, 2020)], poor health outcomes, and ongoing infrastructural disruptions such as power outages, even today. Likewise, in the USVI, many residents have still not recovered their houses, population has not returned, and community organizations and government agencies lack capacity and personnel to work on resilience plans, let alone to implement them. However, there is also another side to this story about how communities came together to help each other.

After the 2017 hurricanes, Puerto Rican activists and ordinary citizens began to challenge the existing modes of climate change policy and governance by taking matters into their own hands. It was clear to all that "community-level actions were fundamental in restoring access to neighborhoods and, ultimately, saving lives" (Rodríguez-Díaz, 2018: 30).

Many community groups began to "rescue" abandoned public schools and other buildings, repurposing them into community resilience hubs. Local volunteers built solar micro-grids to generate their own local energy, and they cooked meals to deliver to those who were housebound. They created community centers, communication hubs, and support for people with physical and mental disabilities. In many cases, these activities developed into longer-term community-led development projects in which "local community members leverage resources (scientific support, lived experience, contextual knowledge, policy language) to advocate for their right to remain (or right to decide) in the face of environmental, social, economic, and political challenges" (Pérez-Quiñones, 2025). Many community-based organizations (CBOs) remain active today in preparing for and responding to extreme weather events, but communication, information sharing, government and academic support, and coordination across efforts remain a challenge.

This book traces the dynamics of climate adaptation efforts within the US Caribbean, with an eye to the wider Caribbean region and how it will "adapt" to climate change in the near future (and whether this is even the right terminology to use). Beyond "natural" causes of climate vulnerability, Caribbean scholars crucially point to the colonial contexts that exacerbate the impacts of climate-induced disasters. Centuries of adverse sociopolitical and environmental issues, shaped by coloniality, have forced people in this region to cope with "unnatural disasters" (Algoed, 2024; García-López, 2018; Rodríguez-Díaz, 2018; Bonilla & LeBrón, 2019). Discourses on and responses to climate change cannot be separated from these longer—and still ongoing—colonial histories and institutional structures that continue to inform present-day vulnerabilities throughout the region (see, for example, Baptiste & Rhiney, 2016; García-López, 2018; Bonilla & LeBrón, 2019; Bonilla, 2020; Sheller, 2018a, 2018b, 2020; Rhiney, 2018; Robinson et al., 2022, 2023). Given this context of coloniality, it is also important for universities and researchers to reflect on their own practices and positions of power, as we shall do in the following chapters.

The US Caribbean archipelagoes are bellwethers of contested federal and local efforts to adapt to climate change, rebuild infrastructure after disasters, and create actionable plans and policies to build greater "resilience" to future climate impacts. The US Caribbean territories' access to US citizenship (although not full citizenship, as some argue), mobility within the United States, and federal funding and resources do not necessarily provide guarantees that advance locally responsive climate adaptation; furthermore, the US Caribbean territories experience comparative disadvantages in terms of being excluded from regional organizations and limited sovereignty over their own policies and regulations (one example is the much-derided Jones Act, which requires US-flagged ships for entry into Puerto Rican ports). Insofar as Caribbean island territories have rightly been problematized as "laboratories" for experiments in "resilience" in the Anthropocene (the era of human-induced climate change), when people are treated as spaces for others to experiment upon (Grove, 2018; Chandler & Pugh, 2021; Sheller, 2020), these experiments play out differently in independent vs. non-independent Caribbean territories in ways that are not fully recognized or understood.

While US Caribbean citizens can move relatively freely to and from the continental United States without a passport, at the same time, residents of US states can freely move to the US territories and in fact have been attracted to do so with targeted tax breaks that in their latest iteration have accelerated displacement and reshaped local economies (Atiles-Osoria, 2025; Jiménez, 2024; Navarro, 2021). This creates a different dynamic of disaster response, climate mobilities, and governance issues, which are unique to these territories. The realities of US citizenship among residents of PR and the USVI therefore do not negate colonial subjugation, or a second-class citizenship status, which is entwined with embodied experiences of gender, racialization, ethnicity, color, class, and sexuality (Negrón-Muntaner, 2018; Navarro, 2021; Meléndez-Badillo, 2024). Research teams that consist of US stateside-based scholars, like ours, must reflect on our privileged access to mobilities and other advantages when we travel to and from the US Caribbean territories, as part of ethical research practices. Beyond access to mobility for researchers, the US Caribbean is also enmeshed in other power dynamics that we seek to reflexively interrogate throughout this book.

Another key concern for climate justice is the ability of people to stay in place. Too often people in PR and USVI have been forced to make the hard choice of migration off-island. Edwin Meléndez and Jennifer Hinojosa (2017) estimated that from 2017 to 2019, "Puerto Rico may lose up to 470,335 residents or 14% of the population," with the largest percentage moving to Florida. Confirming these statistics, in 2020 the US Census Bureau reported significant population losses [-18% in the USVI; −12% in PR] since 2010 (U.S. Census Bureau, 2020). Crises in the United States often impact US Caribbean territories earlier and with more severe consequences. For instance, in her study *Virgin Capital*, Tami Navarro observed how the failed economic policies since the 2007 financial crash, in the form of the Economic Development Commission program (EDC), contributed to the population decline in St Croix. She argues that "the initiative was dogged by much suspicion, stemming from an understanding that the emergent EDC community, comprised of these recently arrived financial managers and their families, was at best snobbish, and at worst racist" (Navarro, 2021, p. 3). Similarly in Puerto Rico, the unelected Financial Oversight and Management Board imposed in 2016 to manage Puerto Rico's bankruptcy, oversee the territory's budget, and ensure access to capital markets has brought harsh austerity measures that have weakened public institutions and led to further outmigration (Cabán, 2018).

Yet, at the same time, the ability to move between the US territories and the states is also a right to free movement that US citizens can exercise as a coping mechanism, which is denied to non-citizens. Non-US citizens—or climate migrants—may be displaced by climate disasters and then just as quickly detained and deported, e.g., as happened to Haitians after Hurricane Dorian in the Bahamas (Nixon, 2019). Historically disadvantaged urban and rural communities in the US Caribbean islands (including Haitian and Dominican migrants) are particularly vulnerable to both long-standing challenges of insecure housing and livelihoods, and the immediate threats that arise with hurricanes, flooding, extreme heat, and recent intensifications of the US deportation regime. The term "climate mobilities" is used to describe these more complex serial displacements which lie at the intersection of class, race, migration, nationality, and the insecurity of citizenship in a climate-pressured world (Boas et al., 2022). Climate mobility justice has emerged as a new concept for better understanding these interlinked relationships of displacement, (im)mobility, and power (Giacomelli & Musarò, 2025).

Meanwhile, emptied-out territories with declining populations are often quickly repopulated by tourists, digital nomads, real estate developers, and others who take advantage of the opportunities of "extreme urbanism" (Dawson, 2017) and the tax structures afforded by the colonial relationship, which in Puerto Rico includes cryptocurrency investors and island-secessionist libertarians (Simpson & Sheller, 2022). The freedom of US citizen investors *to move into* PR or the USVI, supported by economic incentives like EDC in the USVI and Act 60[1] in Puerto Rico, is also a challenge not faced (in the same way) by independent Caribbean nations. One key question guiding this book, then, is what kinds of climate adaptation and "resilience" are actually supporting or limiting these marginalized US Caribbean communities as they seek to resist displacement and remain in place in ways that will hold onto their cultural and social histories, their homes, and land. And, as we explore further in this chapter, what is the role of university-based researchers in relation to these mobility regimes that influence climate adaptation strategies and opportunities?

In the face of recent cutbacks in US federal funding for climate research and climate adaptation, it seems an especially timely moment to reflect on what has been achieved under existing climate adaptation and resilience research initiatives that seek to implement participatory methodologies, and what remains to be done.

3 Overview of the Book

Given the contested context of climate vulnerability, climate adaptation, and resilience planning laid out above, and their relation to uneven (im)mobilities and self-determination, some of the questions this book will grapple with include:

- What are the best approaches—in research and university-community partnerships—to support the "action" work of rebuilding "social infrastructure" after climate disruption, and what limitations does it face? What challenges are faced in conducting community climate adaptation research that bridges academic research objectives and community-based actions, desires, and future orientations?
- What can we learn from those who are practicing co-production of knowledge and participatory planning now (cf. Rodríguez et al., 2024)? How might these collaborations enact co-production that supports multiple ways of knowing and produces knowledge that can

[1] Act 60 is a tax incentive that offers tax exemptions to new residents and export businesses that relocate to Puerto Rico.

directly support communities in their responses to climate change more equitably?

- How can social or human infrastructure for climate justice be implemented at the local level in more equitable and inclusive ways? What would more transformative social infrastructuring do to support climate-adapting communities and what kinds of new possibilities would it lead to?
- To what extent does federal funding such as the NOAA Climate Adaptation Program, and research teams working with shared action learning and community-based participatory methodologies, support actual community efficacy in carrying out planning and preparedness for climate-driven disruptions? Is this even the question that community participants are asking, or the problem they are concerned with? How might researchers reframe climate adaptation in terms that do not assume local priorities and their connections to climate issues?
- What innovative or transformative potentials already exist within frontline communities, and what have we learned from them? And how can new modalities of collaborative climate adaptation, resilient social infrastructure, and creative climate knowledge-action emerge? How can shared action learning, participatory design, and other forms of collaborative research help generate more creative climate adaptation?

Taken together, the chapters will present the overall approach to human-centered design and co-production of the Caribbean Collaborative Action Network. In Chapter 2, the coeditors will introduce the overall research network and its objectives, while highlighting the power dynamics within university-led knowledge-action networks. We recognize the need for sensitivity in questioning our own epistemic privilege as federally-funded researchers, who are able to leave and enter the US Caribbean territories empowered by university credentials.

Chapter 3 focuses on how CCAN implemented human-centered design through undergraduate research projects based on "shared action learning," which is conducted by undergraduate students as part of the WPI Puerto Rico Project Center. The chapter focuses on partnering with a rescued school resilience center in a rural area and examines how trust is built over time even when shaken by unexpected events. It explores how university partners need to remain present, "embedded," and involved over longer time horizons than any single project. Climate resilience and adaptation projects, they show, interact with questions of government and community relations, financing and fundraising, university research timelines, and interpersonal dynamics that can present challenging ethical questions. The authors describe how they navigated one such ethical challenge that demonstrated how co-producing action is a difficult process requiring visibility and trust-building, which ultimately shaped the Project Center's innovative resource-sharing model.

Chapter 4 presents a graduate research project in the US Virgin Islands by two WPI students in the Master's degree in Community Climate Adaptation, supervised by WPI faculty, working with our CCAN partners at the University of the Virgin Islands. Through a community-engaged and participatory research approach, the authors analyze islanders' perceptions of climate risks, vulnerabilities, and resilience as juxtaposed with major climate policy initiatives. Arguing for a decolonial approach to climate adaptation research and community partnerships, the authors highlight the limits to Global North-oriented adaptation planning models and the need for locally grounded solutions that imagine more just, alternative futures for the Virgin Islands. Through this research they laid the groundwork for future ongoing CCAN collaborations in the USVI, while also documenting the challenges and critical positional reflexivity needed to do this work.

Chapter 5 examines CCAN team members' implementation and reflections on the use of the Vulnerability, Consequences, Adaptation Planning Scenarios (VCAPS) methodology in two southern coastal communities in Puerto Rico. Building on a methodology that has been implemented in many community action research situations, Molinari and Tuler identify what they learned about residents' perceptions of climate hazards and vulnerabilities in relation to long-term structural and infrastructural inequalities, and the crucial importance of centering and listening to community voices that frame issues in ways that may differ from planners, policy-makers, and researchers. The chapter shows the limitations of adaptation itself, and how adaptation partnerships and interventions can better support existing social infrastructures—such as women's networks of local action, or fishermen's community-support initiatives—by listening to people's lived experiences and taking seriously local knowledge and connections to place, history, and stories.

Finally, in Chapter 6 we include an additional community-designed project that was not originally affiliated with CCAN, but involves a researcher who is now the Postdoctoral Fellow for CCAN. This chapter focuses on participatory video methodologies that enable people to give voice to their own realities. Focusing on a collaboration with Casa Pueblo in Adjuntas, Puerto Rico, the author explores crucial issues around collaboration, reciprocity, and long-term relationship-building that are essential to transformative education and social action. Like Chapter 3, it also examines trust, accountability, transparency, flexibility, and timing as key to collaborative projects involving academics and community-based organizations. As the author notes, "The timelines and protocols of academia differ from those of communities, making it essential to adapt and navigate these differences to co-produce outcomes that are meaningful and valuable to both parties." And like Chapter 5, it shows how local framings of climate justice matter in ways that can reshape how we engage as researchers. This project suggests some of the future directions we hope to push the CCAN network toward, as we critically reflect on more equitable research relationships and climate-just futures.

Ultimately, in thinking about building greater equity and justice into climate futures, we argue that the emergent qualities of human co-design are relevant to some larger arguments about climate transitions as a kind of "doing together"—that is process-oriented work that is transformative in intention and remains an ongoing unfinished "praxis." We are all engaged in building climate transitions together, while trying to co-produce a more climate-just future. The challenges we describe in this book concern how to achieve that when working especially in Caribbean contexts marked by coloniality; and through universities and federal funding projects that at times remain highly colonial or extractive in terms of their epistemologies and bureaucracies, their uneven spatial relations and the demanding temporalities of research schedules. One approach we suggest is to foster reparative roles for the university to support ongoing local efforts at creating sustainable climate futures, rather than a unidirectional mode of knowledge production or extraction. This means thinking with others about the world we want to create together, and then leveraging university resources and capacities toward this collective objective.

This book synthesizes what the research team has learned through these varied approaches to community-based action research, and where we want to head in the future. Our case studies show that community-centered social (and natural) infrastructures are central to "everyday" or "autonomous" environmental adaptations that people engage with, even in the face of structural limits to adaptation. We consider these collaborative socio-ecological pedagogies as capable of generating more creative climate adaptation approaches. Creative climate adaptation moves beyond expert-led or top-down infrastructural approaches to resilience, yet also transcends some of the current calls for participatory approaches which are sometimes missing a decolonial critique, an analysis of mobile power relations, or an awareness of methods such as *autogestión* (as described in Chapter 6).

In many ways, this is an ongoing journey through the creative "back loops" of Anthropocene experimentation (Wakefield, 2020, 2024). In their everyday practices of people-powered climate adaptation, some local communities in the Caribbean are already occupying public lands and buildings, creating shared physical, social, and natural infrastructures, and building flexible mutual aid collectives that are essential not only in the face of climate disasters but also in response to slower socio-economic pressures, long-term environmental stressors, and ongoing political events. It is this local knowledge and action that we believe is most promising in leading toward greater climate justice and equitable futures.

References

Algoed, L. (2024). *Our Land, Our Survival: Vulnerabilization and Decolonial Resistance through Communal Land in Puerto Rico and Barbuda.* Dissertation.

Atiles-Osoria, J. M. (2025). *Crisis by Design: Emergency Powers and Colonial Legality in Puerto Rico.* Stanford University Press.

Baptiste, A., & Rhiney, K. (2016). Climate Justice and the Caribbean: An introduction. *Geoforum, 73*, 17–21.

Boas, I., Wiegel, H., Farbotko, C., Warner, J., & Sheller, M. (2022). Climate Mobilities: Migration, Im/mobilities and Mobility Regimes in a Changing Climate. *Journal of Ethnic and Migration Studies, 48*(14), 3365–3379.

Bonilla, Y. (2020, April). The Coloniality of Disaster: Race, Empire and the Temporal Logics of Emergency in Puerto Rico, USA. *Political Geography, 78.*

Bonilla, Y., & LeBrón, M. (Eds.). (2019). *Aftershocks of Disaster: Puerto Rico Before and After the Storm.* Haymarket Books.

Cabán, P. (2018). PROMESA, Puerto Rico and the American Empire. *Latino Studies, 16*(2), 161–184.

Chandler, D., & Pugh, J. (2021). Anthropocene Islands: There are Only Islands After the End of the World. *Dialogues in Human Geography, 11*(3). https://doi.org/10.1177/2043820621997018

Dawson, A. (2017). *Extreme Cities: The Perils and Promise of Urban Life in the Age of Climate Change*. Verso.

García-López, G. A. (2018). The Multiple Layers of Environmental Injustice in Contexts of (Un)natural Disasters: The Case of Puerto Rico Post-Hurricane Maria. *Environmental Justice, 11*, 101–108. https://doi.org/10.1089/env.2017.0045

Giacomelli, E., & Musarò, P. L. (2025). *Climate Mobility Justice: Narratives and Visual Politics of the Panicocene*. Palgrave Macmillan.

González, J. E., Georgescu, M., Lemos, M. C., Hosannah, N., & Niyogi, D. (2017). Climate Change's Pulse is in Central America and the Caribbean. *Eos, 98*. https://doi.org/10.1029/2017EO071975. Published on 27 April 2017.

Grove, K. (2018). *Resilience*. Routledge.

Hendricks, M., Meyer, M., & Wilson S. (2022). Moving Up the Ladder in Rising Waters: Community Science in Infrastructure and Hazard Mitigation Planning as a Pathway to Community Control and Flood Disaster Resilience. *Citizen Science: Theory and Practice*. Online first. https://doi.org/10.5334/cstp.462

Jacobs, F. (2019). Black Feminism and Radical Planning: New Directions for Disaster Planning Research. *Planning Theory, 18*(1), 24–39.

Jiménez, M. A. (2024). *Making Never-Never Land: Race and Law in the Creation of Puerto Rico* (1st ed.). University of North Carolina Press.

Makoff, G., & Setser, B. W. (2017). Puerto Rico Update: PROMESA, Population Trends, Risks to the Fiscal and Economic Plan—And Now Maria. *Population Trends, 146*, 28.

McPhillips, L. E., Chang, H., Chester, M. V., Depietri, Y., Friedman, E., Grimm, N. B., Kominoski, J. S., McPhearson, T., Méndez-Lázaro, P., Rosi, E. J., & Shafiei Shiva, J. (2018). Defining extreme events: A cross-disciplinary review. *Earth's Future, 6*, 441–455. https://doi.org/10.1002/2017EF000686

Meléndez-Badillo, J. (2024). *Puerto Rico: A National History*. Princeton University Press.

Meléndez, E., & Hinojosa, J. (2017, October). *Estimates of Post-Hurricane Maria Exodus from Puerto Rico*. Centro. Center for Puerto Rican Studies, Research Brief, Centro RB2017-01. https://centropr.hunter.cuny.edu/app/uploads/2023/03/RB2017-01-POST-MARIA-EXODUS_V3.pdf

Méndez-Lazaro, P. A., Chardón-Maldonado, P., Carrubba, L., Álvarez-Berríos, N., Barreto, M., Bowden, J. H., Crespo-Acevedo, W. I., Diaz, E. L., Gardner, L. S., Gonzalez, G., Guannel, G., Guido, Z., Harmsen, E. W., Leinberger, A. J., McGinley, K., Méndez-Lazaro, P. A., Ortiz, A. P., Pulwarty, R. S.,

Ragster, L. E., Rivera-Collazo, I. C., Santiago, R., Santos-Burgoa, C., & Vila-Biaggi, I. M. (2023). Ch. 23. US Caribbean. In A. R. Crimmins, C. W. Avery, D. R. Easterling, K. E. Kunkel, B. C. Stewart, & T. K. Maycock (Eds.), *Fifth National Climate Assessment*. U.S. Global Change Research Program. https://doi.org/10.7930/NCA5.2023.CH23

Navarro, T. (2021). *Virgin Capital: Race, Gender and Financialization in the U.S. Virgin Islands*. State University of New York Press.

Negrón-Muntaner, F. (2018, January 10). Our Fellow Americans: Why Calling Puerto Ricans 'Americans' Will Not Save Them. *Dissent Magazine*. https://www.dissentmagazine.org/online_articles/our-fellow-americans-puerto-rico-hurricane-maria-colonialism-rhetoric/

Nixon, A. V. (2019, September 9). When the Apocalypse is Now: Climate Crisis, Small Island Disasters and Migration in the Aftermath of Hurricane Dorian. *Stabroek News*.

Pérez-Quiñones, K. (2025). *The Right to Stay: Local Perspectives on the Politics of Disaster, Mitigation Planning, and Habitat Conservation in Espinar, Puerto Rico*. PhD Dissertation, University of Texas, Austin.

Rhiney, K. (2018). Recipe for Resilience? Tracing the Biopolitics of Sint Maarten's Recovery Efforts After Irma. *Journal of Extreme Events, 05*(04), 1840002. https://doi.org/10.1142/S234573761840002X

Robinson, S., Carlson, D., Bouton, E., Dolan, M., Meakem, A., Messe, A., & Roberts, J. T. (2023). The Dynamics of Institutional Arrangements for Climate Change Adaptation in Small Island Developing States in the Atlantic and Indian Oceans. *Sustainability Science, 18*, 251–264. https://doi.org/10.1007/s11625-022-01186-z

Robinson, S., Troncoso, A., Roberts, J., & Peck, M. (2022). "We Are a People": Sovereignty and Disposability in the Context of Puerto Rico's Post-Hurricane Maria Experience. *The Geographical Journal, 189*(4), 575–583. https://doi.org/10.1111/geoj.12472

Rodríguez-Díaz, C. E. (2018). Maria in Puerto Rico: Natural Disaster in a Colonial Archipelago. *American Journal of Public Health, 108*(1), 6–144.

Rodríguez, I., Walter, M., & Temper, L. (Eds.). (2024). *Just Transformations: Grassroots Struggles for Alternative Futures*. Pluto Press.

Sheller, M. (2018a). Caribbean Reconstruction and Climate Justice: Transnational Insurgent Intellectual Networks and post-Hurricane Transformation. *Journal of Extreme Events, 5*(4), 18.

Sheller, M. (2018b). Caribbean Futures in the Offshore Anthropocene: Debt, Disaster, and Duration. *Environment and Planning d: Society and Space, 36*(6), 971–986.

Sheller, M. (2020). *Island Futures: Caribbean Survival in the Anthropocene*. Duke University Press.

Simpson, I., & Sheller, M. (2022). Islands as Interstitial Encrypted Geographies: Making (and Failing) Cryptosecessionist Exits. *Political Geography, 99*, Article 102744.
Stephens, J. C. (2024). *Climate Justice and the University: Shaping a Hopeful Future for All*. Johns Hopkins University Press.
U.S. Census Bureau. (2020).
Wakefield, S. (2020). *Anthropocene Back Loop: Experimentation in Unsafe Operating Space*. Open Humanities Press.
Wakefield, S. (2024). *Miami in the Anthropocene: Rising Seas and Urban Resilience*. University of Minnesota Press.

CHAPTER 2

The Caribbean Collaborative Action Network: Community-Based Climate Action in the US Caribbean

Sarah Molinari and Mimi Sheller

Abstract This chapter introduces the Caribbean Collaborative Action Network (CCAN) and situates it in relation to the evolving activities of community-engaged climate adaptation in Puerto Rico and the US Virgin Islands. Funded under the National Oceanic and Atmospheric Administration (NOAA) Climate Adaptation Partnerships (CAP), CCAN is a multi-pronged knowledge and action network that seeks to support community-based climate adaptation efforts in Puerto Rico and the US Virgin Islands (USVI). The chapter presents the goals of the NOAA CAP

S. Molinari (✉)
Department of Integrative and Global Studies, Worcester Polytechnic Institute, Worcester, MA, USA
e-mail: smolinari@wpi.edu

M. Sheller
The Global School, Worcester Polytechnic Institute, Worcester, MA, USA
e-mail: msheller@wpi.edu

S. Molinari and M. Sheller (eds.), *Building Equitable Climate Adaptation Partnerships in the US Caribbean*,
https://doi.org/10.1007/978-3-032-15727-0_2

program to support community climate adaptation. It then considers critical social science perspectives on "resilience" and adaptation; the concepts of social infrastructuring and the co-production of knowledge; and offers a reflective analysis of the positioning of our own WPI-based research team as part of CCAN. This leads to the concept of creative climate collaborations.

Keywords Participatory action research · Reflexivity · Resilience · Social infrastructure

1 The Caribbean Collaborative Action Network

This chapter introduces the Caribbean Collaborative Action Network (CCAN) and situates it in relation to the evolving activities of community-engaged climate adaptation. Funded under the National Oceanic and Atmospheric Administration (NOAA) Climate Adaptation Partnerships (CAP), CCAN is a multi-pronged knowledge and action network that seeks to support community-based climate adaptation efforts in Puerto Rico and the US Virgin Islands (USVI). CCAN seeks to address these goals through a 5-year project that began in October 2022. CCAN utilizes community-engaged action research strategies to understand and act on local concerns around climate change and climate resilience initiatives with a range of different stakeholders.

The overall aim of the project is to form a regional knowledge-action network of researchers and stakeholders that can evaluate needs, provide technical-scientific expertise, facilitate communication, and build cross-regional connections and climate adaptation capacities from the community micro-level, to the territorial meso-level, to the macro-level of regional networks. The network therefore includes not only local community-based organizations, but also the meso-level agencies and organizations within the two territories, and their connection to agencies and networks such as those of the federal government, and to the wider Caribbean region and international climate adaptation organizations (see Fig. 2.1).

Major CCAN objectives are to: (1) understand how communities respond before, during, and after extreme weather phenomena such as hurricanes and flooding; (2) improve how hazard prediction and planning

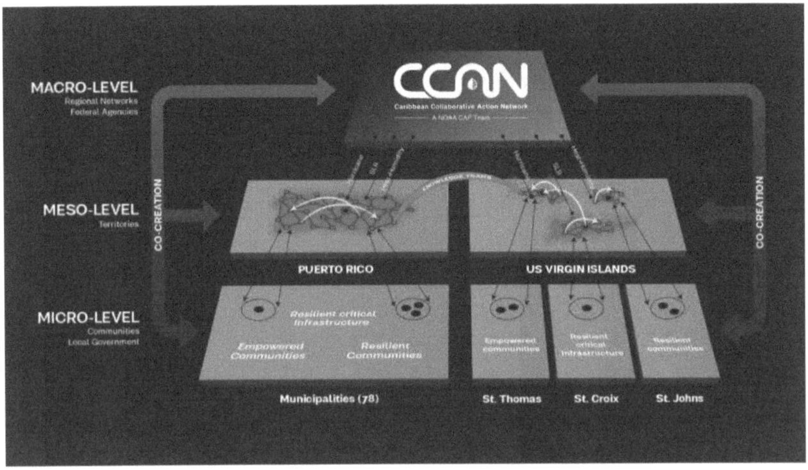

Fig. 1 Schematic overview of the CCAN knowledge-action network

can better reduce local disaster risks; and (3) increase trust, communication, and feedback among scientists, risk forecasters, government agencies, non-governmental organizations, and communities. Beyond that, CCAN also seeks to support communities in their self-defined activities to build "resilience," a problematic term that we return to below. We employ methodologies of co-production and participatory human-centered design with community partners who define the problem, the approach, and the results they want to see. For example, while we began conceiving of the research in the aftermath of hurricanes and landslides that impacted the region over several years, a growing interest in coastal flooding and extreme heat and their effects on human health prompted the CCAN network to pivot to include these topics (see Chapter 5 and Méndez-Lázaro et al., 2025).

Our research team draws on multiple social science and science methodologies that bridge technology and society to gain a deeper understanding of specific places that are affected by climate change, including the local downscaling of climate risks and impacts in ways that are comprehensible to community partners. We also draw on environmental studies and collaboration with engineers and scientists in our CCAN team to gather data, communicate and map risks, and help communities understand complex hydrological and heat systems that are driving the occurrence of extreme heat, drought, flooding, hurricanes, and other local climate-related effects. Working together with communities, governmental and non-governmental organizations, municipal planners, and emergency managers, we seek to analyze climate change risks, situated vulnerabilities, and potential adaptation strategies with community partners.

The CCAN network's human-centered initiatives span workshop facilitation, participatory methodologies, shared action learning projects that bring together undergraduate students, faculty, and local partners, and graduate student research projects. For instance, CCAN researchers have not only built numerous new data collection and analysis tools to measure climate change impacts, but have also hosted policy-engaged workshops on climate change and extreme heat with hundreds of participants from across Puerto Rico, including municipal governments, emergency managers, public school principals, and community-based organizations. An Extreme Heat Summit was also held in the US Virgin Islands in October 2025, involving scientific researchers, community organizations, public agencies, and government officials. In addition to knowledge sharing, CCAN researchers also engage in the co-production of community knowledge and action.

This book focuses especially on the aspect of the work connected with the WPI research team because it includes many of the social scientists involved in human-centered co-design and participatory research. While scientific expertise and quantitative, technical analysis provides insight into the biophysical realities of broad-scale adaptation and future modeling of climate risks, social scientists offer grounded perspectives about the lived experiences of climate risks and the forces that shape the possibilities and limits to adaptation. By centering CCAN's social scientific perspectives in the volume, we highlight how different forms of knowledge production and collaborative actions may shape more just climate futures.

Each chapter in this volume explores different place-based dimensions of adaptation and reflects on the ethics, modes, and challenges within university-community partnerships. WPI's Puerto Rico Project Center has crafted a sustained presence and a shared action learning model linking student research teams and local community resilience centers (see Chapter 3). WPI's Community Climate Adaptation Master's students completed community-based research in the USVI (see Chapter 4). Former WPI postdoctoral fellow Sarah Molinari (2023–2025) and former

UPR doctoral researcher Laura Cabrera worked with community organizations participating in the Whole Community Resilience Planning Program—funded by the US Department of Housing and Urban Development—to lead workshops known as Vulnerability, Consequences, and Adaptation Planning Scenario (VCAPS), which build on a methodology developed by CCAN co-principal investigator Seth Tuler (see Chapter 5). And new WPI postdoctoral fellow Bárbara López-González and collaborators from Casa Pueblo in Adjuntas, Puerto Rico, engage in a methodology known as Participatory Video to show how we can develop more creative approaches to climate justice support networks (see Chapter 6). The contributors will each describe and reflect on the outcomes of this work as we explore the potential of participatory community-centered research to help bring about more just climate futures in the US Caribbean.

In reflecting on our own team's work on this project, we seek not only to assess its value but also to consider how it might best be carried forward in the absence of future federal funding. NOAA CAP's national objectives, according to its mission statement, are to create networks of people working together to support "collaborative research relationships that help communities build lasting and equitable climate resilience" within social contexts. "The CAP program advances NOAA's goal of creating an equitable climate ready nation by working closely with decision makers and frontline communities," according to Wayne Higgins, Director of NOAA's Climate Program Office. In this book, we examine the progress being made in this direction and critically interrogate the process and outcomes of such climate adaptation partnerships at the very moment when funding for them—and climate science itself—is being called into question across the federal government in the United States.

In reflecting on our team's various cases of grounded research, we seek to assess how well a federally funded program that promotes co-production of knowledge and more equitable climate adaptation is actually working. We offer this reflexive analysis at the very moment that such funding is under threat of cancellation if NOAA's research programs are defunded. What is the value of climate adaptation partnerships in a context where science and social science research might struggle to continue? How might these partnerships continue to support frontline communities? Can we leverage university resources toward reparative and justice-oriented engaged research in a context of climate skepticism and shifting federal policies around "diversity, equity, and inclusion?" In this chapter, we will consider critical reflections on the meaning of resilience, creative ways of social infrastructuring, and the research team's reflexive positioning.

2 Beyond Climate "Resilience"

The concept of "resilience" is a common yet contested term in post-disaster and adaptation discourse. Critics argue that it reinforces a neoliberal mindset that exacerbates existing inequalities and individualizes "the resilient self" (Chandler, 2014; Chandler et al., 2020; Grove, 2018; Ooi et al., 2023; Reid, 2018). Others suggest that climate adaptation and resilience policies are associated with gentrification and displacement of poorer communities, deepening existing social and economic fissures (Dryzeck et al., 2013; Erlick, 2021; Taylor & Aalbers, 2022), with outcomes that have especially harmed Black and Afro-Caribbean communities, including in Puerto Rico (Lloréns, 2021). Many historically Black neighborhoods in US cities impacted by hurricanes and flooding were displaced and never able to recover or rebuild affordable housing, for example in post-Hurricane Katrina New Orleans (Johnson, 2011), or more recently in Miami (Grove et al., 2020a, 2020b), and most probably in the wake of the 2025 Los Angeles fires in the historically Black community of Altadena.

Critical perspectives on community climate resilience and climate adaptation initiatives suggest that rather than leading to transformative change, they seek to "manage disruption" to the existing system of inequitable neoliberal urbanism in ways that simultaneously constrain the imaginary of future transformation (Reid, 2018). Even more critically, some scholars argue that we need to move beyond the "coloniality of climate," including contemporary practices of climate adaptation that lead to green gentrification and further displacement of the most vulnerable people (Sheller, 2020a; Sultana, 2024). European colonization, plantation agriculture, and slavery shaped who is most vulnerable to climate disruption in the Caribbean region today—Black, Brown, and Indigenous people, as well as Maroons, fisherfolk, small farmers, and those who migrate to work in other territories. Climate adaptation initiatives need to be inclusive of these communities who bear the brunt of its impacts.

In Puerto Rico's case, "resilience" has circulated in popular discourse especially since Hurricane Maria in 2017, attached to everything from street murals to government funding plans and politicians' celebratory calls for resilience as a personal virtue. However, critics argue that resilience obscures responsibility for harm and normalizes precarity. As scholar Zorimar Rivera Montes writes, "resilience scrambles our understanding of the origins of disaster, making it appear singular and fleeting rather than structural and continual" (Rivera Montes, 2025). Energy infrastructure has been a particular site of struggle in Puerto Rico. For example, in *Revolutionary Power* (2021), energy-justice advocate and lawyer Shalanda Baker "argues that people of color, poor people, and indigenous people must engage in the creation of the new energy system in order to upend the unequal power dynamics of the current system." In *Energy Islands*, Catalina De Onis "provides an urgent and nuanced portrait of collective action that resists racial capitalism, colonialism, and climate disruption." She shows "how local community groups in Puerto Rico have struggled against energy coloniality to mobilize and transform power from the ground up" (De Onis, 2021; see also Dunlap & Tornel, 2023; López-González, forthcoming). These kinds of mobilizations are taking place not only in relation to energy justice, but also housing, education, and food justice.

Many analysts suggest that there are glimmers of experimentation with more transformative practices within everyday climate action and community activism that resist the coloniality of climate and neoliberal ideas of individualized resilience in creative and collective ways (Wakefield, 2020, 2024). Many such transformative practices are being documented in Puerto Rico (e.g., Bonilla & LeBrón, 2019; Lloréns, 2021; López-González, forthcoming; Molinari, 2019; Pérez-Quiñones, 2025; Robinson et al., 2022; Roque et al., 2021). For example, across Puerto Rico communities are creating resilience hubs by "rescuing" abandoned buildings and closed public schools for life-affirming projects (see Chapter 3), defending their right to remain in place and developing actionable asset maps and resilience plans (see Chapter 5), or creating agroecology, solar energy microgrids, and practices of *"autogestión"* (self-management) (see Chapter 6 and, e.g., Bonilla & Lebron, 2019; García-López, 2018). These are the kinds of initiatives we seek to document, collaborate with, and support through CCAN.

In the rural community of Utuado, where some of our CCAN partners supported a community mobilization in the wake of Hurricane Maria, residents continue to expand their resilience hub. Utilizing their own resources, they are pursuing socio-economic development through farm rehabilitation, agritourism and agrotherapy through a project known as

'Tetuan Reborn.'" "This type of self-organization," argue the researchers, "enhances community resilience to disasters. Resilience has implications for sustainability, public health, and risk reduction through higher levels of community cohesiveness, social capital and coordination. Finally, the Utuado model of resilience could extend out to the region and the entire island" (Holladay et al., 2019, p. 5), for example through the development of community-based tourism that is sustainable and supports local livelihoods, enabling people to stay in their home region (Holladay et al., 2021; cf. Sheller, 2020a, 2020b).

In St. Thomas, USVI, organizations such as We Grow Food Inc. have created long-term agroecology projects that defend both the land and culture, providing Virgin Islanders with affordable, locally grown produce and direct assistance to farmers. These efforts support collective needs, build social solidarity, and offer solutions and adaptations that demonstrate how climate stressors are embedded within and inseparable from socio-political milieus. But they can only go so far, without wider political, economic, and ecological transformations. In the face of extreme storms like Hurricane Melissa, which in October 2025 grew into a slow-moving, category five tropical cyclone, devastating parts of southern Haiti, western Jamaica, southeastern Cuba, and nearby islands, it will take years if not decades to recover. As the chapters in this book will consider, what are the possibilities and limits of different strategies for climate adaptation? What is the relation between local initiatives, local government support (or its absence), federal policies, and research and educational initiatives to support climate adaptation? And where and when does accountability take hold for those who are most responsible for the greenhouse gas emissions that are over-heating the planet?

3 Social Infrastructuring

The governance of climate adaptation has largely shifted from a top-down control model that once focused on rebuilding physical infrastructure, to a distributed and participatory logic of community-based resilience (Wakefield, 2024). As urban sociologist Eric Klinenberg argued after the 1995 Chicago heatwave, and after Hurricane Sandy in 2012 New York, it was just as important to pay attention to "social infrastructure" and rebuild connections among people and the public places that support such connectivity (Klinenberg, 2015). Co-design, community empowerment, inclusive design, and researcher-practitioner partnerships have become the watchwords in the fields of climate adaptation and resilience (Dryzek et al., 2013; Erlick, 2021; Rajkovich & Holmes, 2022; Shandas & Hellman, 2022). During the Obama and Biden administrations, it was commonly posited by federal funding agencies, emergency managers, urban planners, architects, engineers, and post-disaster researchers that the challenge of climate adaptation requires human-centered design and participatory action (Saum-Manning, 2021; Wakefield, 2024).

Whether one remains critical of the piecemeal management of state withdrawal, or remains hopeful about transformative decentralization, within both perspectives is the question of strengthening social infrastructure. Studies of disaster recovery have demonstrated the importance of "human infrastructure" (Klinenberg, 2015, 2018), mutual aid (Villarrubia-Mendoza & Vélez-Vélez, 2020), and the building of "infrastructural commons" (Frischmann, 2013) as crucial parts of disaster response and mitigation. Many regions of the United States, but especially the US territories like Puerto Rico, are facing failures and deliberate dismantling of their physical *and* social infrastructure. While much attention has been given to the failure of the electricity grid, the closure of hospitals and schools, and the lack of water during extreme heat and droughts, equal attention must be given to the social infrastructure of people who keep things together and social care systems functional. Outside experts, engineers, or organizations have no "participants" to work with unless there is existing social infrastructure or their research activities are actively designed to build or strengthen social infrastructure.

The question we want to pose in relation to this literature is to ask how research projects by university faculty and students like those of CCAN can be integrated into the practice of social *infrastructuring*—as an active practice with potentially transformative outcomes and creative initiatives. How can higher education research teams help to support social infrastructure through this challenging period? Are we directing our research, classwork, student projects, and collaborations toward the development of solutions to real-world problems (Montalto, 2023)? How successful are we at building trust by co-producing knowledge with communities (Feagan et al., 2025), while using multiple methods to bring people together and reach the most marginalized (Saum-Manning, 2021)? More specifically, we are interested in the ways that academic researchers, federal funding agencies, and local communities are trying to work together toward improving community climate adaptation through involving "stakeholders" in meaningful participation and co-production that goes beyond simply sharing information or gaining buy-in. We are interested in social creativity in the face of climate change and the opportunities that education and research offer.

The notion of "co-production of knowledge" has become especially salient, drawing on the idea of human settlements as social-ecological-technological systems, or SETS (Feagan et al., 2025), and participatory pathways toward climate adaptation through the "the co-production of sustainable future scenarios" (Iwaniec et al., 2020). The activities of the CCAN network, as described in each of the chapters below, involve collaborative research activities that are not only supporting various kinds of social infrastructure for climate adaptation, but are trying to co-design it together with participants. To some extent, this offers a creative and emergent methodology for building social infrastructure as a knowledge-action network that comes out of the practice of direct collaboration. The purposeful integration and sharing of different kinds of community-generated knowledge and action can support the capacity of multilayered local agency in ways that may be especially crucial now, as formal federal support for climate-related projects dries up. At least, that is the hope of the network and something that we are trying to assess as we move forward.

On the other hand, there are also many limitations and power dynamics involved in the research approaches that we are engaged in. This requires a strong dose of reflexivity about our own positionality to consider who brings what knowledge to the table and what kinds of knowledge are valorized in different ways. It necessitates that we deeply reflect on whether we are contributing to the mere management of climate disruption (a bandaid covering deeper structural inequities) or if we are able to build relationships through community-engaged research that contributes to the social transformation that some are advocating for. Even "stakeholder participation" and the strengthening of social infrastructure can in practice uphold harmful forms of neoliberal "resilient" development that still displace marginalized communities or fail to meet their basic needs (with many Caribbean communities experiencing ongoing power and water disruptions, food insecurity, flooding, extreme heat, closed schools, lack of healthcare access, etc., long after immediate emergencies are over).

While advocating the merits of some kinds of participatory approaches, we want to acknowledge that they have both strengths and weaknesses. As Montalto and Sheller argue, "Social infrastructure will continue to function with or without federal funding and policy. Indeed, it is the way countless marginalized communities have combatted oppression for centuries" (Montalto & Sheller, 2025). When the communities we work with through CCAN build relationships with our research teams, they are not only seeking to adapt to climate hazards but are also challenging us to adapt the structural limitations of the current system to better meet their needs and support their visions of justice, including epistemic justice. The questions we pose about the CCAN project seek to reflexively consider in what ways it does or does not contribute to building and strengthening social infrastructure that can lead toward more equitable processes and just outcomes for communities experiencing climate disruption, extreme weather, and environmental stressors, especially within non-sovereign or colonized territories.

4 Reflexive Positioning of WPI-CCAN Research Activities

Worcester Polytechnic Institute (WPI), a science and engineering university established in 1865 in central Massachusetts, has employed a project-based curriculum since the 1970s. The teaching mission supports inquiry-based learning, open-ended problem solving, and integrative and interdisciplinary thinking. One facet of this is the junior year project known as an Interactive Qualifying Project (IQP), an interdisciplinary requirement involving applied research that connects science or technology with social issues and human needs. Students work with local sponsoring organizations to address complex local and regional problems over a period of 14 weeks, with the final seven weeks in residence at one of WPI's 50 project centers around the world. The Puerto Rico Project Center (PRPC), currently directed by WPI faculty members John-Michael Davis and Scott Jiusto (who are also part of the CCAN team; see Chapter 3), is one of WPI's longest continuously operating project centers, having hosted groups of students every year since 1992. Based in San Juan, students work in interdisciplinary teams with sponsoring organizations both in the metropolitan area and elsewhere around the archipelago.

The PRPC supports undergraduate student teams to work on research projects focusing on sustainable community development and includes a diverse portfolio of themes related to community climate resilience, agroecology, and environmental and emergency management.

In conjunction with the emergence of the CCAN network, the PRPC was able to expand the scope, geographic locations, and interconnections of the projects. Our program is innovative among academic institutions for the breadth and depth of commitment to grassroots community groups in PR. Our projects focus on topics determined by community partners and advanced in close collaboration with them. This kind of grounded research can not only seek solutions to the challenges such groups face in addressing issues of community/climate resiliency, but also can engage deeply in the shared effort to develop, test, revise, and share practical strategies, tools, and resources for meeting challenges of community adaptation and well-being. This led to an effort in the Spring of 2025 to launch a new Community Resilience Platform, that seeks to build from the student projects to offer community groups online tools and resources and ongoing support to promote well-being and climate resilience in their communities (see Chapter 3).

WPI also offers a Master's degree in Community Climate Adaptation, and CCAN has been able to support several students to undertake research in PR and the USVI (see Chapter 4). Along with the work of CCAN Postdoctoral Fellows based in the region, this has added additional dimensions to expanding the knowledge-action network. Furthermore, CCAN is partnering with the Consortium for Climate Risk in the Urban Northeast (CCRUN), through an additional WPI-led NOAA-BIL grant, on a study seeking to share, document, and improve methods for community collaboration, citizen science participation, and building equitable climate adaptation strategies.[1] CCRUN is supported by NOAA's Climate Adaptation Partnerships (CAP) program and conducts stakeholder-driven research in the urban Northeast United States. Their team has a strong network of decision makers, local and municipal officials, and research and subject matter experts, along the corridor running from Philadelphia to New York City, and to Boston, a region many Caribbean diaspora communities call home.

These research efforts build on regional and diasporic academic networks across the US and wider Caribbean region. For instance, the Center for Puerto Rican Studies at Hunter College is the largest university-based research institute, library, and archive dedicated to the

[1] NOAA-BIL Grant #NA23OAR4310396-T1-01, *Improving Engagement Methods for Coastal Resilience and Reducing Climate Risk: Bridging Learning Networks From the Urban Northeast (CCRUN) to the US Caribbean (CCAN)*. PI, Mimi Sheller, WPI.

Puerto Rican experience in the United States (https://centropr.hunter.cuny.edu/). The Virgin Islands Studies Collective (VISCO) is "a group of academics, artists, and activists who are committed to centering the Virgin Islands as a site of inquiry and theorization beyond a notion of utopia or space that is not meaningfully occupied" (https://www.vistudiescollective.org/). CCAN researchers seek to build bridges with these regional and diasporic networks, as well as with the wider Caribbean region where other international initiatives are working on climate adaptation, claims for climate reparations, and policies for greater climate justice. The Greater Caribbean Climate Mobility Initiative, for example, is led by the Global Centre for Climate Mobility and the Association of Caribbean States in close collaboration with UN partners and the World Bank. It "aims to build a shared evidence base on climate mobility and to chart a common course of action" centered on inclusive knowledge, the right to stay, documenting and mitigating loss and damage, international solidarity with the region, and if planned relocation is necessary then its basis should be "climate mobility in dignity" and local inclusion in decision making (Global Center for Climate Mobility, 2025).

In reflecting on our CCAN initiatives, we first want to note the positionalities in these institutional arrangements. Most of the WPI research team, and most of the students whom we supervise, are not from the Caribbean region. As outsiders, our positionalities, including racial, ethnic, color, linguistic, gender, and sexual identities, shape our approach to and engagement with Puerto Rican and Virgin Islander realities. Most of the team is not based in the region for extended periods of time. One postdoctoral researcher is from Puerto Rico, and the other had extensive ties and lived there for several years; Masters students were based in the USVI for a four-month semester; undergraduate students are in Puerto Rico for a seven-week term with the Puerto Rico Project Center; and other researchers visit occasionally. Some of us have long-standing connections, affinities, and commitments with the Caribbean through research, kin, and activism, while others are newly engaging with the region through carefully embedding relationships. Many of us do not speak Spanish, which meant that certain research activities were reserved for fluent Spanish speakers.

Secondly, the Caribbean-based team members in CCAN, at University of Puerto Rico and University of the Virgin Islands, come more from the physical sciences than the social sciences. These dynamics of academic disciplines have both opened interdisciplinary knowledge production and positioned some of the emphasis and local network resources more within a quantitative and scientific analysis of Caribbean climate risks, vulnerabilities, and adaptation. Beyond these disciplinary differences, there are also other differences in capacity. We learned from colleagues at the University of the Virgin Islands that there were significantly fewer resources and organizations that could be drawn into the "knowledge-action network" that CCAN sought to mobilize. The USVI has a more horizontal governmental structure, with no municipalities, and fewer public sector employees and limited staffing in governmental offices. There are also far fewer community-based organizations. Thus it is more challenging for the social science researchers to find partners or community organizations to work with in USVI, while it may also be challenging for scientific data to be translated into public policy in directly actionable ways.

Thirdly, we also want to note that academic mobilities and multi-sited research are always shaped by the physical and social infrastructure of getting to places (including costs, time, effort) and the uneven geographies and mobilities of the US Caribbean archipelagos, with their own dynamics of centers and peripheries. For instance, Puerto Rico's island municipalities of Vieques and Culebra are marginalized in relation to the "big island," and St. Thomas figures as the center of resources in the USVI in relation to St. John and St. Croix. Virgin Islanders as a whole also experience marginalization in the US Caribbean archipelagos in relation to Puerto Rico, with its comparatively larger government and resource infrastructures. The coloniality of these archipelagic geographies of (im)mobility shapes our knowledge production, its limitations, and future directions. That is why "archipelagic thinking," which grows out of the work of Caribbean theorists such as Antonio Benítez Rojo, Derek Walcott, and Edouard Glissant can be a powerful tool for decentering more privileged spaces and geographies, and recentering the lives and perspectives of islanders (Miguel & Stephens, 2020; Sheller, 2020c). For instance, CCAN itself is trying to expand its archipelagic focus by moving forward in an effort to "recenter" the margins, with more resources committed to the "peripheral" islands. A key question is, what insights can be gleaned about climate adaptation from the archipelagic margins?

The CCAN research team has had to wrestle with all of these issues of academic positionalities, disciplinary divisions, archipelagic epistemologies, and research (im)mobilities while thinking about how they intersect with "climate mobilities." Differently positioned islanders may or may not have access to various kinds of mobilities—such as migration to the continental United States—and may face displacement and rising costs due to collapsing infrastructure and the arrival of outsiders who are buying up property in desirable island locations, or renting AirBnB's and enjoying higher disposable incomes—including academic researchers. With all of these limitations in mind, we nevertheless believe it is worthwhile to try to conduct university-community partnerships like CCAN that stretch across institutions, geographies, and different kinds of resources. One way that social scientists propose for addressing the limitations of outside research interventions is through "co-production" of knowledge. Drawing on lessons from feminist scholarship, knowledge co-production encompasses participatory research, action research, and engaged scholarship that attempts to unsettle hierarchical modes of knowledge production while recognizing multiple ways of knowing and doing. Academic researchers engage in co-production by bridging gaps with various stakeholders, centering situated knowledges, and problematizing the assumption of epistemic authority. Co-production posits an "active immersion in, rather than distant, critical observation of, messy and complex life-worlds" (Scheba & Scheba, 2023, 2; see also Patel, 2022). As we collectively confront the conditions of the climate crisis, some scholars argue that co-production processes are more urgent than ever (Norström et al., 2020).

In methodologies of Participatory Action Research (PAR), "local knowledge forms the basis for research and planning" (Cornwall & Jewkes, 1995, p. 1667) in ways that can catalyze collective action and reconfigure social relationships (Santiago Ortiz, 2020). PAR involves researchers and stakeholders "in a process through which they are empowered to take charge of the research process and to organize to implement potential solutions or to take action" (Cornwall & Jewkes, 1995, p. 1671). Hendricks et al. argue that robust participatory action research, feminist research, and community-based citizen science can foster the co-production of knowledge in ways that empower marginalized groups and recognize differing forms of valuation and risk assessment. The key element in all these approaches is the integration of social knowledge and dialogical co-learning processes into community understandings of risk, adaptation, and transformation, so that people can take action themselves. However, this is not always easy for university-based researchers enmeshed in highly bureaucratic institutions and federal funding requirements to achieve in reality. What is called for now is the assessment of different methodologies for equitable participation in real-world contexts.

The chapters in this volume work through co-production of knowledge and participatory research methodologies in different ways, including shared action learning approaches that bring together faculty, students, and community partners (Chapter 3), community-engaged participatory research (Chapter 4), reflexively adapting to the challenges of participatory methodologies (Chapter 5), and co-designed creative investigations with community collaborators (Chapter 6). These examples move through an arc that extends from supervised undergraduate research, to graduate-level fieldwork, to faculty and postdoctoral research, to deeply co-designed community co-production. It also moves from members of our team who are less grounded in Puerto Rican or Virgin Islands contexts, to those who are most deeply connected to these places. Thus we are able to explore different kinds of collaboration, different positionalities, and different lessons learned. We hope this will be valuable to other students, researchers, teachers, community organizers, and research collaborators, as they also embark on collaborations toward more climate-just futures.

5 Conclusion: Toward Creative Climate Collaborations

The chapters gathered in this book point us toward some lessons about the reflexive co-production of what we call "creative climate collaborations" as sets of ongoing relationships that are co-evolving and dynamic, rather than quick-fix technical solutions or simple models of "climate resilience" as bouncing back to a prior state. Climate creativity can be understood, in broad terms, as everyday practices grounded in people's efforts of bricolage and appropriation that lash things together in new ways. Creative actions to deal with a changing climate are emergent improvisations. These practices are provisional, dynamic, and go through ups and downs depending on who is present and able to contribute, what resources are available, and what specific needs various communities identify as important at different times. This offers a new way to think about climate adaptation that goes beyond resilience by seeking out positive community transformations.

Every day people join together to make things work, to envision a future of thriving communities where residents can stay in place, and to bring possibilities for improvement into their lives—whether it's getting the lights to stay on, making an emergency plan for their community, delivering meals to the elderly, or supporting fisherfolk in times of hardship. These are all examples of creative climate collaborations that emerge out of everyday climate adaptation activities. Insofar as research activities can identify, document, and support such creative activities, our collaborations with communities can become a site of creative climate adaptation that can leverage resources for reparative ends.

In the process of building such social infrastructure, many vulnerable communities also learn to diagnose the structural inequities that are preventing them from thriving, whether it is tourists and vacation property owners moving in, or government failures to fund water, energy, and sewer infrastructure. They challenge academic research to move beyond "objective" science that simply measures problems and reports data, to consider instead how education can be extended to participatory action research and creative collaborations that support community-led initiatives.

This presents an opportunity to try to build more just climate futures through our collaborative action research. The emergent qualities of human-centered co-design mean that we are building climate transitions together. Climate transition—or better yet transformation—calls for repairing social infrastructures as much as physical infrastructures, and that includes changing the ways that university-based researchers interact with community-based organizations as collaborators in the co-production of knowledge and climate action.

The challenge remains how we can do that in a context of climate coloniality, characterized by government abandonment, ongoing extraction, and climate skepticism. What does climate adaptation mean for territories of the US Caribbean that are non-independent, and face ongoing calls for decolonization? How do social scientists and other researchers concerned with climate impacts engage with the politics of decolonizing knowledge when our knowledge systems, university institutional structures, and federal funding remain highly colonial in many ways? Climate justice calls on us to identify historical patterns of inequity, to rebalance unequal positionalities, and to recognize ongoing challenges of coloniality (which are not always easy to uproot). The research presented here seeks to show how we are trying to build ethical research relationships, which we believe are crucial to advancing the project of just climate adaptation.

Acknowledgements We want to thank all of our collaborators in CCAN who have supported this work and helped make it possible. Led by Principal Investigator Pablo Méndez-Lázaro, the full list of members can be found at http://ccan-upr.org. We thank the many community organizations who have partnered with the WPI Puerto Rico Project Center (https://wp.wpi.edu/puertorico/) during the course of this research. We also thank our colleagues in the Global School at WPI, including the dedicated faculty and staff involved in the Global Projects Program, and operations management and administrative assistance from Dawn Farmer and Nancy Fay.

Competing Interests This study was partially supported through the Caribbean Collaborative Action Network (CCAN), a NOAA Climate Adaptation Partnerships team. Grant Number NA22OAR4310545 (PI, Pablo Méndez-Lázaro). It was also supported in part by the NOAA-BIL Grant Number NA23OAR4310396-T1-01, Improving Engagement Methods for Coastal Resilience and Reducing Climate Risk: Bridging Learning Networks From the Urban Northeast (CCRUN) to the US Caribbean (CCAN) (PI, Mimi Sheller, WPI).

References

Baker, S. (2021). *Revolutionary Power: An Activist's Guide to the Energy Transition*. Island Press.

Bonilla, Y., & Lebron, M. (Eds.). (2019). *Aftershocks of Disaster: Puerto Rico Before and After the Storm*. Haymarket Books.

Chandler, D. (2014). *Resilience: The Governance of Complexity*. Routledge.

Chandler, D., Grove, K., & Wakefield, S. (Eds.). (2020). *Resilience in the Anthropocene: Governance and Politics at the End of the World*. Routledge.

Chandler, D., & Pugh, J. (2021). Anthropocene Islands: There are Only Islands After the End of the World. *Dialogues in Human Geography, 11*(3) https://doi.org/10.1177/2043820621997018

Cornwall, A., & Jewkes, R. (1995). What Is Participatory Research? *Social Science & Medicine, 41*(12), 1667–1676.

De Onis, C. 2021. *Energy Island (Environmental Communication, Power, and Culture)* (1st ed., Vol. 1). University of California Press.

Dryzek, J. S., Norgaard, R. B., & Schlosberg, D. (2013). *Climate-Challenged Society*. Oxford University Press.

Dunlap, A., & Tornel, C. (2023). Pluralizing Energy Justice? Towards Cultivating an Unruly, Autonomous and Insurrectionary Research Agenda. *Energy Research & Social Science, 103*, Article 103217.

Erlick, J. C. (2021). *Natural Disasters in Latin America and the Caribbean: Coping with Calamity*. Routledge.

Feagan, M., Muñoz-Erickson, T. A., Hobbins, R., et al. (2025). Co-Producing New Knowledge Systems for Resilient and Just Coastal Cities: A Social-Ecological-Technological Systems Framework for Data Visualization. *Cities, 156*, Article 105513.

Frischmann, B. M. (2013). *Infrastructure: The Social Value of Shared Resources*. Oxford University Press.

García-López, G. A. 2018. The Multiple Layers of Environmental Injustice in Contexts of (Un)natural Disasters: The Case of Puerto Rico Post-hurricane Maria. *Environmental Justice, 11*, 101–108. https://doi.org/10.1089/env.2017.0045

Global Center for Climate Mobility. (2025). *The Greater Caribbean Climate Mobility Report: Agenda for Action*. https://greatercaribbean.climatemobility.org/agenda-for-action.

Grove, K. (2018). *Resilience*. Routledge.

Grove, K., Barnett, A., & Cox, S. (2020a). Designing Justice? Race and the Limits of Recognition in Greater Miami Resilience Planning. *Geoforum, 117*, 134–142.

Grove, K., Cox, S., & Barnett, A. (2020b). Racializing Resilience: Assemblage, Critique, and Contested Futures in Greater Miami Resilience Planning. *Annals of the Association of American Geographers, 110*(5), 1613–1630.

Holladay, P. J., Méndez-Lázaro, P., Brundiers, K. (2021). From Hurricanes to Pandemics: Community-Based Transformation and Destination Resilience in Utuado, Puerto Rico. *Journal of Sustainability and Resilience, 1*(2), Article 1. https://digitalcommons.usf.edu/jsr/vol1/iss2/1

Holladay, P. J., Méndez-Lázaro, P., Centeno, H. M., Rivera-Gutierrez, R., Adams, K., & Brundiers, K. (2019). Utuado, Puerto Rico and Community Resilience Post-Hurricane Maria: The Case of Tetuan Reborn. *Recreation, Parks, and Tourism in Public Health, 3*, 5–19.

Iwaniec, D. M., Cook, E., Davidson, M., et al. (2020). The Co-production of Sustainable Future Scenarios. *Landscape and Urban Planning, 197*, Article 103744.

Johnson, C. (2011). *The Neoliberal Deluge: Hurricane Katrina, Late Capitalism, and the Remaking of New Orleans*. University of Minnesota Press.

Klinenberg, E. (2013, October). Toward a Stronger Social Infrastructure: A conversation with Eric Klinenberg. *Urban Omnibus, 16*. https://urbanomnibus.net/2013/10/toward-a-stronger-social-infrastructure-a-conversation-with-eric-klinenberg/

Klinenberg, E. (2015). *Heat Wave: A Social Autopsy of Disaster in Chicago*. University of Chicago Press.

Klinenberg, E. (2018). *Palaces for the People: How Social Infrastructure Can Help Fight Inequality, Polarization and the Decline of Civic Life*. Penguin Random House.

Lloréns, H. (2021). *Making Livable Worlds: Afro-Puerto Rican Women Building Environmental Justice*. University of Washington Press.
López-González, B. (forthcoming). Art for Re-Existence: Casa Pueblo's Decolonial Landscape Interventions Toward an Energy Insurrection. *Human Geography*.
Martínez-San Miguel, Y., & Stephens, M. (Eds.). (2020). *Contemporary Archipelagic Thinking: Toward New Comparative Methodologies and Disciplinary Formations*. Rowman & Littlefield.
Méndez-Lázaro, P. A., Cabrera-Rivera, L. T., Rueda-Roa, D., et al. (2025). Assessing the Impacts, Risks, and Vulnerabilities of Extreme Heat in Learning Environments of Puerto Rico in 2023. *The Journal of Climate Change and Health, 26*, Article 100581. https://doi.org/10.1016/j.joclim.2025.100581
Molinari, S. (2019). Authenticating Loss and Contesting Recovery: FEMA and the Politics of Colonial Disaster Management. In Y. Bonilla & M. LeBrón (Eds.), *Aftershocks of Disaster: Puerto Rico Before and After the Storm*. Haymarket Books.
Montalto, F. A. (2023). How Tackling Real-World Problems Transformed my Teaching and Research. *Nature, 621*, 659. https://doi.org/10.1038/d41586-023-02989-5
Montalto, F. A., & Sheller, M. (2025, June 4). Strengthening Social Infrastructure for Climate Resilience. *The Nature of Cities*. www.thenatureofcities.com/TNOC/2025/06/04/strengthening-social-infrastructure-for-climate-resilience/
Montes, Z. R. (2025, April 30). *It's About Time We Fail: A Puerto Rican Poetics of Exhaustion. Public Books*. Retrieved 22 July 2025: https://www.publicbooks.org/its-about-time-we-fail-a-puerto-rican-poetics-of-exhaustion/
Norström, A. V., Cvitanovic, C., Löf, M. F., et al. (2020). Principles for Knowledge Co-Production in Sustainability Research. *Nature Sustainability, 3*(3), 3. https://doi.org/10.1038/s41893-019-0448-2
Ooi, Can-Seng, et al. (2023). *Islands and resilience: Experiences from the Pandemic Era*. Springer.
Santiago Ortiz, A. (2020, December). Mapping Collaboration as Neoliberal Resistance in Participatory Action Research in Puerto Rico *Tracce Urbane. Rivista Italiana Transdisciplinare Di Studi Urbani, 8*, 269–289. https://doi.org/10.13133/2532-6562_4.8.17220
Patel, Z. (2022). The Potential and Pitfalls of Co-Producing Urban Knowledge: Rethinking Spaces of Engagement. *Methodological Innovations, 15*(3), 374–386. https://doi.org/10.1177/20597991221129779
Pérez-Quiñones, K. (2025). *The Right to Stay: Local Perspectives on the Politics of Disaster, Mitigation Planning, and Habitat Conservation in Espinar, Puerto Rico* (PhD Dissertation). University of Texas, Austin.

Rajkovich, N. B., & Holmes, S. H. (Eds.). (2022). *Climate Adaptation and Resilience Across Scales: From Buildings to Cities*. Routledge.
Reid, J. (2018). Securing the Imagination: The Politics of the Resilient Self. In J. Bohland, S. Davoudi, & J. Lawrence (Eds.), *The Resilience Machine* (pp. 29–42). Routledge.
Robinson, S., Troncoso, A., Roberts, J., & Peck, M. (2022). "We are a People": Sovereignty and Disposability in the Context of Puerto Rico's Post-hurricane Maria experience. *The Geographical Journal*. https://doi.org/10.1111/geoj.12472
Roque, A., Wutich, A., Brewis, A., Beresford, M., García-Quijano, C., Lloréns, H., & Jepson, W. (2021). *Autogestión* and Water Sharing Networks in Puerto Rico After Hurricane María. *Water International, 46*(6), 938–955. https://doi.org/10.1080/02508060.2021.1960103
Saum-Manning, L. (2021). Best Practices and Lessons Learned from Community Engagement and Data Collection Strategies in Post-Hurricane Maria Puerto Rico. *Journal of Homeland Security and Emergency Management, 18*(3), 225–250. https://doi.org/10.1515/jhsem-2020-0075
Scheba, S., & Scheba, A. (2023, May 30). Co-producing Just Geographies: Resourcing, Bridging, and Critical Crossings in Engaged Scholarship. *Dialogues in Human Geography*. https://doi.org/10.1177/20438206231178810
Shandas, V., & Hellman, D. (Eds.). (2022). *Collaborating for Climate Equity: Researcher-Practitioner Partnerships in the Americas*. Routledge.
Sheller, M. (2020a). *Island Futures: Caribbean Survival in the Anthropocene*. Duke University Press.
Sheller, M. (2020b). Reconstructing Tourism in the Caribbean: Connecting Pandemic Recovery, Climate Resilience, and Sustainable Tourism Through Mobility Justice. *Journal of Sustainable Tourism, 29*(9), 1436–1449. https://doi.org/10.1080/09669582.2020.1791141
Sheller, M. (2020c). Archipelagoes as the Fractal Fringe of Coloniality: Demilitarizing Caribbean and Pacific Islands. In Y. Martínez-San Miguel & M. Stephens (Eds.), *Contemporary Archipelagic Thinking* (pp. 287–306). Rowman & Littlefield.
Sultana, F. (Ed.). (2024). *Confronting Climate Coloniality: Decolonizing Pathways for Climate Justice*. Routledge.
Taylor, Z., & Aalbers, M. (2022). Climate Gentrification: Risk, Rent and Restructuring in Greater Miami. *Annals of the Association of American Geographers, 12*(6), 1685–1701.
Villarrubia-Mendoza, J., & Vélez-Vélez, R. (2020). Centros de Apoyo Mutuo: Reconfigurando La Asistencia En Tiempos de Desastre. *Centro Journal, xxxii*(iii).

Wakefield, S. (2020). *Anthropocene Back Loop: Experimentation in Unsafe Operating Space*. Open Humanities Press.
Wakefield, S. (2024). *Miami in the Anthropocene: Rising Seas and Urban Resilience*. University of Minnesota Press.

CHAPTER 3

Supporting Social Infrastructure in Puerto Rico Through Shared Action Learning

John-Michael Davis and Scott Jiusto

Abstract This chapter explores a five-year university-community partnership in Puerto Rico that supported the transformation of an abandoned school into a community resilience hub. Through Worcester Polytechnic Institute's Puerto Rico Project Center and its Shared Action Learning approach, faculty and students collaborated with Id Shaliah, a grassroots organization in the rural community of Cubuy-Lomas, Canóvanas, to co-produce knowledge and action that advanced climate resilience. The chapter situates this work within broader challenges facing Puerto Rico, including widespread school closures, aging populations, and heightened community vulnerabilities to disasters. Through three vignettes, we illustrate how sustained engagement, institutional networks, and tangible

J.-M. Davis (✉) · S. Jiusto
Department of Integrative and Global Studies, The Global School, Worcester Polytechnic Institute, Worcester, MA, USA
e-mail: jdavis4@wpi.edu

S. Jiusto
e-mail: sjiusto@wpi.edu

S. Molinari and M. Sheller (eds.), *Building Equitable Climate Adaptation Partnerships in the US Caribbean*,
https://doi.org/10.1007/978-3-032-15727-0_3

contributions foster reciprocal relationships and strengthen the social infrastructure for community-defined futures. We argue that universities, when engaged in long-term, trust-based partnerships, can play a catalytic role in co-producing both physical and social infrastructure by supporting grassroots initiatives with visibility, resources, and solidarity. The chapter concludes by reflecting on the evolution of the Puerto Rico Project Center into a hub for rescued school partnerships and the development of a Community Resilience Platform to amplify local expertise and foster shared learning. This work contributes to the literature on ethical global learning, community-based research, and climate adaptation by demonstrating how long-term, trust-based collaborations can support marginalized communities to envision and enact resilient futures.

Keywords Community resilience · Shared action learning · University-community partnerships · Climate adaptation · Rescued schools

1 Serendipitous Beginnings

Sitting in the backseat of an Uber, we began the 45-minute drive from San Juan, Puerto Rico, to the rural community of Cubuy-Lomas, Canóvanas, in search of an abandoned school. It was early September 2020; we traveled from our Massachusetts university, Worcester Polytechnic Institute (WPI), in search of sponsor organizations interested in hosting students for our faculty-led project-based learning program. While WPI has operated an education program in Puerto Rico for over 30 years that sends students to work on projects with local organizations, we had just taken over as program directors, were new to Puerto Rico, and eager to start networking and building relationships with new partners.

Before our arrival in Puerto Rico, we conducted some initial internet research to identify and contact dozens of organizations we thought might be interested in partnering with us. Karma Honey Project, an environmental nonprofit organization whose mission was to "save the bees," responded to our e-mail. We arranged to meet with Johnny, an organizational representative, and his associate, who was affiliated with a San Juan-based university, during one of our first afternoons in Puerto Rico. We waited at an outdoor bar with large picnic tables on Calle Loiza, a rapidly gentrifying region of San Juan. Johnny arrived on a motorcycle, was in his mid-40s, fast-talking and charismatic, and seemingly well-connected throughout Puerto Rico. We discussed several project possibilities with Karma Honey Project. One that struck our attention was their recent efforts to repurpose an abandoned school into a community resilience center with a community-based organization (CBO) in Cubuy-Lomas, 45 minutes away. We thought it would be a good idea to see the project for ourselves.

As we drove closer to Cubuy-Lomas, we began traversing up the narrow, winding mountainous road on the outskirts of El Yunque rainforest. It became increasingly clear how vulnerable these rural communities can be. Landslides, fallen trees, or even a car accident can block access roads and obstruct evacuation routes for hours, if not days. After Hurricane Maria, wealthier urban communities had electricity restored within a few months whereas many rural areas, like Cubuy-Lomas, were left in the dark for seven months or longer. Beyond its geographic vulnerability, the majority of the 10,000-person population is aged 65 or older with an average employment rate of 32% and a median income of $18,780 (United States Census Bureau, 2021). The Cubuy-Lomas community was rather typical of rural regions throughout Puerto Rico.

After a few unplanned detours, we arrived in Cubuy-Lomas at the abandoned Manuel Agosto Lebrón School, which served pre-K to grade 5 and closed two years prior in 2018. The school sits atop a mountainside with views of El Yunque rainforest juxtaposed with the eerily tragic remnants of school paraphernalia strewn throughout the property. Classrooms with faded educational materials on the walls, a basketball court with a crooked net and cracked concrete, overgrown weeds everywhere—telltale signs of neglect and abandonment (Fig. 1).

We reunited with Johnny and met his local partner, Javier, who had similar charismatic, fast-moving energy. Javier grew up in Cubuy-Lomas and established a CBO, Id Shaliah, during the COVID-19 pandemic to operate several programs to support residents, including a meal delivery program, medical supplies deliveries and doctor visits to bedridden residents, and a home rescue program to renovate damaged homes. As we toured the school, Javier and Johnny shared their plans to create a community center and resilience hub. They envisioned a medical clinic and pharmacy in the principal's office, a community kitchen where the school cafeteria used to be, a shelter with beds and canned food in one classroom, a small apiculture business incubator in another, kick-boxing

Fig. 1 Our first visit to the Manuel Agosto Lebrón School in Cubuy-Lomas in the Fall of 2020

classes upstairs, and so on. All the hallmarks of effective community resilience that involve both the ability to respond to and recover from disasters, but also proactive measures to foster social cohesion, economic stability, environmental sustainability, and overall well-being (Cutter et al., 2008). Yet, they had made little progress so far and shared the challenges they encountered: slow-moving bureaucratic government processes and difficulty securing funding, resources, volunteers, and a sustainable operational model. They shared ambitious goals to open the school by early 2021, just a few months away. That ended up being overly ambitious by about three years.

This chapter highlights our partnership with Id Shaliah over a five-year period that supported their efforts to transform the abandoned school into the Cubuy-Lomas Centro de Desarrollo and Servicios Comunitarios (Community Development and Services Center). We describe how school closures and increasing elderly populations throughout Puerto Rico have complicated community resilience in climate-vulnerable regions. We then situate WPI's Global Projects Program (GPP) and our program, the Puerto Rico Project Center (PRPC), to discuss how we foster university-community partnerships through a Shared Action Learning (SAL) approach with grassroots groups and CBOs. Through three brief vignettes, we illustrate key insights that led our partnership with Id Shaliah to evolve into a deeply impactful, reciprocal relationship and how it catalyzed the reimagination of our program in Puerto Rico to better support community-led climate resilience.

2 Puerto Rico's Abandoned Schools

The abandonment of the Manuel Agosto Lebrón school was part of a broader restructuring of Puerto Rico's public education system. Since 2007, Puerto Rico has closed nearly 44% of all public schools in the archipelago (Yedidia et al., 2020). These closures have left hundreds of school buildings vacant, creating a void where lively educational centers once thrived and posing a complex regional development challenge regarding the future use of these abandoned properties. School closures have been driven by a mix of emigration, natural disasters, and austerity measures in response to Puerto Rico's debt crisis. The mass closures began in 2007, primarily due to increased emigration (Hinojosa et al., 2019; Yedidia et al., 2020). The population of individuals under 18 in Puerto Rico has declined from approximately 1,018,000 in 2006 to 597,000 in 2021 (US Census Bureau, n.d.). In 2017, the Puerto Rican government declared bankruptcy and the Oversight Board[1] pursued school closures, which accelerated following Hurricanes Irma and Maria. Overall, some 650 schools closed across Puerto Rico.

In rural areas like Cubuy-Lomas, school closures have profound consequences that extend beyond the loss of educational facilities. Schools often serve as vital community hubs, hosting health clinics, daycare centers, after-school programs, recreational spaces, and emergency shelters. They act as "third places" where residents build relationships and supportive networks, contributing to a strong sense of community identity (Rosenbaum et al., 2019; Witten et al., 2001). When schools close, this sense of identity can deteriorate, and neighborhoods become less attractive to families, encouraging young families to relocate and potentially threaten the economic viability of the entire community (Cranston, 2017;

[1] The Financial Oversight and Management Board for Puerto Rico (commonly referred to as the "Oversight Board" or "La Junta") was established by the U.S. Congress under the 2016 PROMESA Act to oversee Puerto Rico's fiscal recovery and debt restructuring. While intended to stabilize Puerto Rico's finances, the Oversight Board has been widely criticized for imposing austerity measures, including mass public school closures.

People for Education, 2009). Moreover, abandoned school buildings can decrease property values, invite vandalism, discourage investments, and impose financial costs on municipalities, ultimately lowering residents' quality of life (Cohen, 2001). In Puerto Rico, school closures have led to longer commutes for families, declining student enrollments, and deteriorating academic performance (Rivera Rivera, 2022; Rodríguez, 2015; Yedidia et al., 2020). The remaining open schools often face overcrowding, with some requesting FEMA trailers to accommodate the influx of students in the wake of the 2017 hurricanes (Martinez & Lee, 2018). Taken together, these closures have fractured long-standing community resources and weakened community resilience.

Puerto Rico is now grappling with the challenge of determining the future of its abandoned schools. Some aspire to sell these properties to generate revenue and help alleviate Puerto Rico's debt crisis. Others hope to see these schools repurposed to benefit society. In 2017, Governor Ricardo Rosselló signed an executive order to address the ownership of abandoned properties by establishing a committee chaired by the Fiscal Agency and Financial Advisory Authority (AAFAF). This committee, along with the Evaluation and Transfer of Real Estate Assets Committee (CEBDI), was tasked with creating a system for efficiently evaluating proposals for renting or selling school properties. The vision for these closed schools was to support activities for the common good, whether through nonprofit or commercial organizations, or by boosting the real estate market and economy. The bureaucratic process, however, has been criticized for a lack of transparency and for making it difficult for grassroots community groups, which often lack the necessary status designations and financial resources, to submit applications and gain access to these properties.

Despite the government's vision for closed schools to "engage in activities for the common good" (Executive Order No. 2017-032), most remain vacant and communities aspiring to revitalize a lost asset face major bureaucratic and operational challenges. At the same time, grassroots organizations throughout Puerto Rico have taken the initiative to "rescue" abandoned school buildings, often without government permission, and repurpose them as vital community resources (Davis et al., 2023). Through persistent negotiations with government agencies, these groups have managed to occupy, lease, or, in rare cases, gain ownership of vacant schools. They provide a variety of programs and services that address local development priorities, maintain the role of schools as "third places," and work to repair residents' relationship with their community identity and heritage.

In 2023, a WPI student research team in collaboration with Taller Comunidad la Goyco, a rescued school project in San Juan, identified 38 such rescued schools throughout Puerto Rico providing a wide range of services, including arts and cultural programming, community kitchens and food banks, disaster relief and resilience hubs, job creation opportunities, medical clinics, community libraries, environmental programs, and psycho-social and educational services for children (Davis et al., 2023). Rescued schools serve as community anchor institutions, particularly in rural areas, by addressing challenges associated with aging populations, the out-migration of young people, and emergency preparedness. For example, Cubuy-Lomas is located 30–40 minutes from the nearest hospital and Id Shaliah aspired to provide basic medical services such as a pharmacy and access to physician care at their rescued school. Many rescued schools incorporate community resilience resources as a key function of their projects, offering access to solar power, potable water, food, and other emergency preparedness resources. These initiatives not only envision and enact community futures but also address localized issues, demonstrating the potential of rescued schools to foster community resilience and development.

Despite the restorative value that rescued schools offer, repurposing them is a significant undertaking, especially when they have been vacant for years. One of the primary challenges community initiatives face is legally acquiring school properties. Competing with more lucrative bids, the Puerto Rican government often prioritizes cost savings from school closures. The Department of Transportation and Public Works (DTOP), which holds the power to sell State property that has ceased to be of public use, typically asks for around $500,000 to purchase abandoned schools (a figure multiple community groups have independently shared with us), making ownership unattainable for most grassroots organizations. Communication with DTOP and understanding competing interests in vacant schools is often opaque, and lease applications can take several years to approve. Many community groups feel a sense of ownership due to their ongoing maintenance efforts and productive use of the abandoned spaces, advocating for the schools to be returned to the commons. Through lengthy rounds of communication and advocacy, rescued schools are often granted short-term leases for 1–3 years at a nominal fee, but without long-term leases or transfer of title, they remain in a precarious position to continually demonstrate their value while facing the threat of new entities offering lucrative bids at the end of their lease.

Obtaining and sustaining sufficient funding is another major challenge for rescued schools. Volunteer labor and in-kind contributions from within the community are the main resources supporting these initiatives. While rescued schools have achieved varying success in securing donations from CBOs, Puerto Rican foundations, NGOs, private businesses, and contributions from the broader Puerto Rican diaspora, these tend to be ad hoc and short-term arrangements. Municipal governments have occasionally provided nominal support, such as covering electricity and water costs or providing construction materials for renovations, but these are inconsistent. Consequently, rescued schools often progress as fast as available funding permits, typically renovating and repurposing one room at a time. Sustaining community buy-in and volunteer support is also a significant challenge. Initial efforts to clean debris and rubbish from abandoned schools attract enthusiasm and support from the community, but the ongoing physical labor, coupled with the need for engineering, legal, and management expertise, makes mobilizing and coordinating volunteers a substantial undertaking. Rescued schools operating solely with ad hoc volunteers are in a precarious position. Thus, university partnerships can be particularly valuable to sustain the herculean efforts of rescued schools as they can provide resources, expertise, and the enthusiasm of students eager to contribute to an inspirational cause.

3 Supporting University-Community Partnerships Through the WPI Plan

Our collaboration with Id Shaliah was only possible through WPI's institutional commitments to radically embrace project-based learning and global experiential opportunities for all students. WPI is a science and engineering university established in 1865 in central Massachusetts and has employed a project-based curriculum since the 1970s. Known as the "WPI Plan," this groundbreaking shift in undergraduate education transformed a traditional technological curriculum into a student-centered, project-based approach to learning that is designed to prepare students for real-world challenges by emphasizing hands-on, experiential learning.

The WPI curriculum requires students to complete an Interactive Qualifying Project (IQP)—an interdisciplinary project that addresses the intersections of science, technology, and societal needs. The IQP is designed to achieve multi-faceted, transformative learning experiences that develop student skills related to professionalism, teamwork, writing, cultural competency, critical thinking, and problem-solving, while producing real-world contributions that positively impact community partners. Today, the GPP includes 50+ project centers in the United States and abroad. The PRPC is one of WPI's longest operating project centers, having hosted student groups since 1992. The increased institutional support and student participation in global projects propelled our reimagination of the PRPC when we began directing the project center in 2020, and our inclusion as part of the Caribbean Collaborative Action Network (CCAN), a NOAA Climate Adaptation Partnership team that began in 2023, further supported this work. Since then, we've evolved the PRPC into a dynamic hub for community-based research and action to provide transformative learning opportunities for students while meaningfully contributing to the missions of our project partners and CCAN. We operate two cohorts of students annually. Each cohort comprises about 24 students (six teams of four junior-year students), along with two faculty advisors, who collaborate on projects identified and sponsored by Puerto Rican organizations. From the student perspective, the program involves three phases: a summer reading session to learn about contemporary issues in Puerto Rico, a two-month intensive preparation course at WPI to develop research skills and a project proposal for their sponsor before arriving in Puerto Rico, and a two-month full-time research project in Puerto Rico. This structure ensures that students are well prepared and deeply immersed in their projects and in the Puerto Rican context, fostering both academic and practical growth.

Our collaborative work through the PRPC has employed a "Shared Action Learning" (SAL) approach that engages students, faculty, and Puerto Rican partners in projects where all parties learn and benefit from one another to support sustainable development in marginalized communities (Jiusto et al., 2013). SAL emphasizes Sharing among partners of ideas, knowledge, resources, inspiration, and compassion; Action that supports the creative impulses and growth of communities and partners; and Learning drawn from research, action, and critical, reflexive practice.

Like our encounter with Johnny and Javier, projects typically emerge through faculty networking with potential sponsors—organizations and agencies interested in having a team of students advance a topic of concern. The initial phase involves establishing a working relationship between faculty and sponsors, sharing risks and limitations, and setting expectations. Students then take on the responsibility of strategizing and conducting research to respond effectively to the project challenge. Faculty serve as advisors, helping students understand how to advance their projects and develop topical expertise. This preparation phase includes developing a formal project proposal that summarizes background research and proposed plans for field activities.

Community engagement becomes a critical component of SAL once the IQP term begins, immersing students in complex learning environments and often leading to a redefinition of the project based on input from key stakeholders. This process teaches students to navigate unforeseen barriers and opportunities, fostering resilience and adaptability. The action phase involves implementing project plans, often requiring students to build networks with professional and community partners. The subsequent reflection and reporting phases involve students preparing various forms of documentation and presentations to distill their experiences and propose future actions, while critically reflecting on their learning experiences.

As we demonstrate in the following sections, we've prioritized partnering with CBOs in Puerto Rico, recognizing them as key actors to improve climate resilience. CBOs play a vital role in promoting health and well-being in communities, especially in the face of increasing extreme weather events and disasters. For example, studies in the wake of Hurricane María have criticized federal and local disaster response and recovery agencies for their slow and inadequate response (Milman, 2017). An underappreciated reality of disasters is that community members tend to serve as first responders (Cull, 2019), and in Puerto Rico, CBOs become critical actors to mobilize social capital for resource allocation including food, energy, and information networks (Roque et al., 2020). These organizations provide essential rescue and relief support during emergencies and help communities cope with the aftermath. Despite their crucial role, CBOs often receive insufficient support and resources to meet community needs. In what follows, we elaborate on our sustained partnership with Id Shaliah to demonstrate the reciprocal value and lessons learned from our university-community partnership.

4 Building Resilience Together

Our initial visit to Cubuy led to two student projects that began in the Fall of 2021. These projects spurred a five-year, and still ongoing, partnership that combined transformative student learning experiences with targeted projects that advanced Id Shaliah's rescued school aspirations, leading to its official opening in 2024. Our experiences working with Id Shaliah made clear the important role universities can play toward building resilience in climate-vulnerable communities. Traditional university-community partnerships often position university partners as "knowledge generators," producing research and data that, hopefully, inform decision-making. While this has a valuable role, it has also been criticized for reinforcing one-directional flows of expertise and overlooking the lived knowledge and agency of community members (Strier, 2014; Weerts, 2005). Our SAL approach navigates this dynamic by emphasizing mutual learning, co-creation, and sustained engagement. Through this model, we came to understand that communities often value university partnerships not just for technical expertise, but for the tangible support, visibility, and solidarity they can offer in navigating complex social and physical infrastructural challenges.

In what follows, we reflect on our five-year collaboration with Id Shaliah through three vignettes that illustrate how university-community partnerships can evolve into deeply impactful, reciprocal relationships. Each vignette explores a distinct but interconnected theme that emerged through our SAL approach: the resilience fostered through long-term commitment, the catalytic potential of embedded institutional networks, and the value of co-producing action that meets immediate community needs. These stories not only highlight the practical outcomes of our work together but also offer insights into how universities can more equitably and effectively support community-led climate resilience.

4.1 *Sustaining Partnerships Through Crisis: Relationships as Social Infrastructure*

Our partnership with Id Shaliah began in 2021 with two projects supporting their vision for the Cubuy-Lomas community center and resilience hub. Together, these early projects helped articulate and amplify the center's mission by producing a vision statement, website, 360-degree virtual tour, and a planning toolkit, while also advancing infrastructure planning through a resiliency guide focused on solar energy and potable water systems, which advanced the conceptual and physical groundwork for future development and deepened trust between our teams (Brennan et al., 2021; Conway et al., 2021). Moreover, the students benefited from the hands-on learning experience and participation in a bottom-up, community-driven effort to enhance climate resiliency, all while developing meaningful relationships with a receptive and welcoming community.

In the months following our initial projects, we were blindsided by a startling development that challenged the integrity of our partnership and raised difficult questions about due diligence in community-based research. Johnny, who initially helped facilitate our connection with Id Shaliah, was revealed to have misrepresented his identity and was arrested, and then extradited to Canada in connection with an ongoing investigation and criminal charges. Johnny had been a fugitive in Puerto Rico for many years and had done well to obscure his identity. Despite his history, he was known as a philanthropist in Puerto Rico, ran an environmental NGO to help bees thrive, held fundraising events to buy Christmas toys for children, and facilitated an extensive network of volunteers. He was well-connected too, maintaining relationships with politicians, wealthy residents, higher education institutions, government agencies, CBOs, and us academics. For the Cubuy-Lomas center project, Johnny was a potentially transformative partner. His arrest was big news in Puerto Rico. Institutions affiliated with him were shaken.

In the aftermath of the arrest, Javier and Id Shaliah came under intense scrutiny. Although Javier had been hurt and deceived by Johnny's betrayal, their prior association led many to question his integrity, like whether donations to Id Shaliah were actually supporting the community center. Many of Id Shaliah's allies, including volunteers, financial supporters, and organizations that provided resources for the rescued school, distanced themselves. The rescued school project stalled, and Javier was left to defend his reputation while trying to keep Id Shaliah afloat. While less consequential, we received questions from university lawyers, administrators, and parents for our community partnership choices. These latter questions were ameliorated by the fact that we were hardly alone in having been unaware of Johnny's background and had based our relationship largely on Javier's standing with the community and the municipality, which also supported our collaboration. Despite the complications, we remained in close contact with Javier, offering support and solidarity. Together, we decided to refrain from student projects the following year but made it clear that we were not abandoning Id Shaliah. That continued presence, both institutional and personal, became a gesture of solidarity and a stabilizing force that Javier has indicated was highly appreciated as he rebuilt credibility and signaled to others that Id Shaliah remained a trusted and valued partner.

By 2023, our partnership had resumed with renewed purpose, and Id Shaliah was once again positioned to lead resilience-building efforts in Cubuy-Lomas. One of the clearest signs of how much our collaboration meant to Javier came when he and one of his core volunteers traveled to Massachusetts to meet with our incoming student team for a week before their project began. This visit was unprecedented—not just for our project center, but across WPI's entire GPP, which includes over 50 international and domestic sites. Sponsors rarely make such a journey. Their visit reflected a deep commitment to the partnership and a desire to build personal connections with the students who would soon be working alongside them. It also marked a turning point: The bad press had faded, and Javier had begun to rebuild relationships with the local community, the municipality, and former supporters.

This experience underscored how long-term, trust-based university-community partnerships can support grassroots organizations navigating crises and how such relationships are essential to sustaining the social infrastructure of climate resilience. Social infrastructure refers to the networks, institutions, and relationships that enable communities to organize, share knowledge, and care for one another—especially in times of disruption (Montalto & Sheller, 2025). In Cubuy-Lomas, this social infrastructure was embodied in the enduring relationships, mutual care, and shared commitment that kept Id Shaliah afloat during a period of intense scrutiny. As Montalto and Sheller (2025) argue, "robust social infrastructure is one of the most important foundations for resiliency," because it enables the integration of technical expertise with local knowledge and fosters adaptive capacity through mutual support and deliberation. In our case, this infrastructure laid the groundwork for tangible resilience-building outcomes, such as the installation of solar panels described below, demonstrating how social ties can catalyze material improvements and connect communities to broader networks of support.

This experience also reminded us that both academic and community partners have legitimate concerns about who they associate with, and that such concerns often only emerge and can be addressed through the process of working together. While rare and extreme, situations like this underscore the importance of transparency, reflexivity, and sustained engagement. Fortunately, our experience with partners in Puerto Rico and elsewhere has been overwhelmingly positive, and even in this unusual case, our commitment to mutual support helped the partnership endure and grow stronger.

4.2 From Vision to Voltage: Leveraging Embedded Networks for Community Resilience

One of the most underappreciated contributions universities can make to community partnerships is their ability to serve as network facilitators by connecting grassroots organizations to broader ecosystems of knowledge, visibility, and resources. This role is especially valuable for small CBOs that often operate with limited capacity and visibility. This connective function is built into the PRPC's organizational structure. Each year, we partner with about twelve community organizations across the archipelago, creating opportunities for knowledge exchange, resource sharing, solidarity building, and collaborative problem-solving. The "Interactive" element of the IQP often takes the form of students networking with various actors and stakeholders. This capacity for network-building was especially evident in our early work with Id Shaliah. In 2021, student teams not only explored water and power resilience strategies for the rescued school, but also developed a suite of visibility tools to help the organization communicate its mission. As part of their research, students also engaged in informal networking on behalf of Id Shaliah to share their story with other NGOs, funders, academic partners, and other potential allies. One of the most significant outcomes of this visibility-building work came two years later, when the University of Wisconsin-Madison Engineers Without Borders (EWB) chapter contacted us regarding a solar panel system installation. They had discovered Id Shaliah through the website created by our students and the PRPC website, and were interested in donating and installing a solar panel system at the rescued school to increase community energy resilience.

The University of Wisconsin-Madison EWB chapter had originally planned to install a $60,000 solar energy system at a local church in the same Cubuy-Lomas community. However, a leadership change within the church led to an abrupt cancellation of the project. Recognizing the opportunity, we worked closely with the EWB team and a 2023 WPI student team to facilitate a new connection with Id Shaliah (Holmes et al., 2023). The timing was ideal: Id Shaliah had recently resumed its development efforts and was actively seeking sustainable energy solutions for the rescued school. Through this collaboration, the solar panel system was redirected and installed at Id Shaliah's rescued school in January 2025 (Fig. 2). This outcome was not the result of a single project, but rather the cumulative effect of sustained engagement, strategic visibility, and the PRPC's embedded presence in the community.

This story illustrates how universities can serve as connective networks to enhance community resilience efforts. By helping partners like Id Shaliah articulate their vision and amplify their presence, the PRPC played a catalytic role in expanding their visibility and attracting new allies. In this case, a student-built website and a series of collaborative projects became the bridge between a small rural community in Puerto Rico and a major infrastructure investment from across the continent. When engaged in long-term, trust-based partnerships, universities can leverage their institutional credibility and visibility to open opportunities that build not only physical infrastructure but also enhance the social infrastructure essential for climate resilience and community empowerment. At the same time, success is never guaranteed, and together we often miss the mark, witness

Fig. 2 A section of the solar panel installation at Id Shaliah's rescued school

plans for an emergency water system, fundraising strategies, engineering assistance, etc., that remain unrealized.

4.3 Co-producing Action: Material Contributions to Community Resilience

There has been growing emphasis on the co-production of knowledge in university-community partnerships, which represents a shift away from traditional, top-down models where universities act as sole knowledge producers and communities are recipients. These conventional approaches have been critiqued for reinforcing power imbalances and overlooking the lived expertise of community members. In response, more equitable models have emerged that center community voices in shaping research agendas, defining questions, and participating in the generation of knowledge. While this evolution is both necessary and promising, our experience suggests that many communities value something even more immediate action. Through our work with Id Shaliah, we found that co-producing action through tangible, hands-on contributions presented a different set of challenges but was far more meaningful than only co-producing knowledge as an end goal. These visible, practical outputs not only addressed urgent needs but also built trust, legitimacy, and organizational momentum.

Across our five-year collaboration with Id Shaliah, nearly every student project included some tangible output whether it was central to the project or offered as a service contribution. These ranged from painting classrooms and cleaning debris to developing blueprints, building a community garden, and creating digital tools. These physical and functional contributions helped legitimize the partnership in the eyes of the community and demonstrated that student work could translate into meaningful, on-the-ground progress. They also reflected a deeper ethic of reciprocity: that university partners should not only learn from and with communities but also leave behind something useful, visible, and valued. These contributions were publicly recognized during the 2024 opening ceremony, where community members highlighted not only the sustained support of student teams but pointed to the visible impact of their work (Fig. 3).

Id Shaliah's launch of a novel clothing distribution program in 2024 called Project Salvage demonstrated the unique benefits and challenges associated with co-producing action to improve the operational capacity

Fig. 3 Grand opening ceremony of Id Shaliah's Centro de Desarrollo & Servicios Comunitarios in February 2024

of community-led initiatives. Project Salvage emerged from a creative partnership Javier facilitated with a major retail outlet in Puerto Rico to receive unsold merchandise for redistribution. The model required Id Shaliah to donate 80% of the merchandise received to nonprofit organizations across the archipelago and sell the rest to support their rescued school through a thrift boutique called "Armario." To support this initiative, a WPI student team developed a set of tools and systems to improve the program's efficiency and sustainability. One of Javier's top priorities was a digital system to track outgoing donations. The student team conducted interviews with thrift stores and Id Shaliah volunteers and determined that a barcode-based inventory system and a more robust network of donation recipients would be the most appropriate solutions to improve operations. They developed a barcode scanner system along with an automated Google Form and spreadsheet database for uploading donation images, and a survey that expanded Id Shaliah's merchandise distribution network to over 40 nonprofit partners (Duggan et al., 2024).

Yet, after the students left, Javier struggled to operate the barcode scanner and the system went unused for months. It wasn't until the following year, during a new student project, that we learned the system had never been fully implemented. When we inquired whether this reflected its usefulness or relevance, Javier shared that he truly wanted to use the system but couldn't figure out how to make it work. In response, we arranged a virtual meeting between Javier and the student team that developed the system to troubleshoot and walk through its use. With that support, the system was finally activated and is now functioning as intended. This experience underscored a critical lesson: Co-producing action requires more than delivering a product, and it requires sustained relationships, iterative learning, and a commitment to follow-through. In this way, our partnership with Id Shaliah revealed that the most impactful university-community collaborations are not just about generating knowledge or building tools, but about staying in the work long enough to ensure those tools are actually implemented and serve the people they were designed for.

5 Reimagining the PRPC as a Network for Community Resilience

Our partnership with Id Shaliah became a catalyst for the PRPC to support rescued schools as key intervention points to enhance community resilience in Puerto Rico. While WPI partners with a wide variety of organizations including government agencies, professionalized NGOs, businesses, museums, universities, and so on, in Puerto Rico, we focus on CBOs who often benefit disproportionately from the sustained support of a university partner and the periodic surge of youthful energy from inspired student researchers. A sustained (though not exclusive) focus on rescued schools also helped us learn that most rescued schools face similar challenges related to securing ownership of their school, managing volunteers, developing programs, and identifying sustainable funding sources, which presented synergistic opportunities for our program to facilitate resource and knowledge sharing. Moreover, rescued schools offered transformative learning opportunities for our students to understand the challenges of bottom-up community resilience and to be inspired by their altruistic efforts and generosity. Through a combination of internet research and conversations with allies, we expanded our network of rescued school partners and positioned our program as a network connector by facilitating resource and knowledge exchanges and co-creating tools that amplify the efforts of grassroots organizations driving climate resilience.

For example, we began our partnership with Taller Comunidad la Goyco in 2022, a rescued school transformed into a community center in Santurce, San Juan. La Goyco has become a vibrant hub for social services, cultural activities, health initiatives, and environmental projects for their community. Compared to other rescued schools, La Goyco is highly advanced in their development with full-time staff, a completely renovated school, and a sustainable funding model based on grants, citizen donations, and income from renting classrooms to local businesses. Our collaboration has included projects that supported the creation of a rescued school network (Abrogar et al., 2023), documented the local impacts of gentrification (Govani et al., 2023), and developed volunteer management tools (Chesanek et al., 2023), an emergency preparedness toolkit (Fallon et al., 2023; Hasbrouck et al., 2024; Li et al., 2025), and a citizen science noise monitoring system (Blanchard et al., 2024). We also benefited from our collaboration with La Goyco by renting classrooms and community spaces to provide office space for students and a place to host weekly cohort meals and participate in cultural immersion opportunities. We then expanded further. We learned about the Centro de Apoyo Mutuo Las Carolinas (CAM-LC) in Caguas, a CBO that emerged in 2017 when residents and allies reclaimed an elementary school to initiate a community kitchen. The CAM-LC then became a hub for local emergency response, food distributions, and socialization. From 2024 to 2025, we collaborated on three projects that (a) developed an emergency response system that created a community census protocol to identify assets and needs, (b) identified resources the CAM-LC has and what is needed to prepare a resilience hub, and (c) developed an initial plan to respond during a disaster (Checko et al., 2024; Davis et al., 2025; Diaz et al., 2024; Gagnon et al., 2025). We also partnered with La Conde, a community project working to transform the six-acre Carlos Conde Marín School campus in Carolina into a creative neighborhood space, where WPI teams developed digital volunteer and inventory tools in 2024 and designed and piloted a community garden with irrigation, composting, and social programming features in 2025 (Ambrad et al., 2025; Kawatkar et al., 2024). We continued learning about and partnering with other groups scattered throughout Puerto Rico and have since facilitated partnerships with eight rescued schools.

The structure of our program has enabled us to facilitate a rescued school network through a kind of spoke-and-hub engagement model to share resources and knowledge among grassroots groups that would otherwise have minimal opportunities to learn from each other. For example, we introduced Id Shaliah to CAM-LC to provide guidance on how to navigate government relationships to obtain long-term leases and eventual ownership of school buildings, and through Project Salvage, Id Shaliah donated clothing to distribute to the Las Carolinas community. Moreover, projects completed with specific rescued schools have broader relevance. Emergency management plans and volunteer management systems can be adapted and applied in other contexts. We have also secured external grant resources to support our rescued school network. For example, in 2025 we were awarded a Sociological Initiatives Foundation grant to clarify government processes and advocacy strategies for school ownership transfer and increase the political influence of communities working to repurpose abandoned schools into community centers.

Our experiences with Id Shaliah and other community partnerships made clear the value of a sustained presence, which led us to expand our organizational infrastructure within Puerto Rico. Most notably, we hired two Puerto Rican scholar activists as PRPC coordinators and advisors who bring extensive expertise in community organizing, rescued spaces, and climate resiliency. Their lived experience and local knowledge have been instrumental in bridging cultural and logistic gaps to enhance the continuity and cultural relevance of our work.

Together, we are developing the Community Resilience Platform, a resource website for grassroots community groups across Puerto Rico to access online tools and resources and direct support to promote climate resilience. The platform focuses on four key areas: emergency management, food sovereignty, rescued community spaces, and the operational needs of small CBOs. Beyond providing resources, the platform fosters mutual support among community groups and amplifies the expertise of local partners, creating a digital commons for shared learning and collective action. With funding from the Caribbean Collaborative Action Network (CCAN), we hired a platform coordinator to maintain year-round engagement with community groups, ensuring that our partnerships are not limited to student project cycles—and that when things we create stop working (like Id Shalah's barcode scanner), we have more capacity to respond quickly.

Our work with Id Shaliah and other rescued school initiatives in Puerto Rico contributes to a growing body of scholarship on community-based participatory action research, which emphasizes justice-oriented, place-based, and collaborative approaches to knowledge production. For decades, PAR scholars and practitioners have insisted that research should not only interpret the world but also help change it. Our SAL model exemplifies this ethos by integrating sustained engagement, mutual learning, and tangible contributions that support community-defined goals. Our partnership with Id Shaliah has demonstrated how universities can provide solidarity, visibility, and resources to support grassroots organizations navigate complex challenges and enhance resiliency. Importantly, our experience underscores the value of co-producing not only knowledge but also action. Community research that fails to generate tangible improvements risks reinforcing extractive dynamics; by contrast, our SAL approach affirms that communities often value practical contributions as much as, if not more than, academic insights.

We understand the social infrastructure for climate resilience as the networks, relationships, and institutions that enable communities to organize, adapt, and care for one another in the face of disruption. This infrastructure is not merely supportive to physical infrastructure—it is foundational. Our chapter demonstrates how university-community partnerships, when rooted in trust and shared purpose, can help build and sustain both physical and social infrastructure with grassroots organizations at the front of climate resilience efforts, thereby enhancing community adaptive capacity and fostering more equitable and enduring climate resilience. In this way, our work contributes to a growing recognition in the literature (e.g., Montalto & Sheller, 2025) that technical solutions alone are insufficient—resilience must be cultivated through the social systems that make those solutions viable and meaningful.

Competing Interests This study was partially supported through the Caribbean Collaborative Action Network (CCAN), a NOAA Climate Adaptation Partnerships team. Grant Number NA22OAR4310545.

Ethics Approval Ethics approval was granted by the Ethics Committee of Worcester Polytechnic Institute. Informed consent to participate and to publish was obtained from all individuals involved in the study.

References

Abrogar, J., Bourgoin, J., Brown, M., Kellum, E., & Polito, F. (2023). *Facilitating a Network of Rescued Schools in Puerto Rico.* Interactive Qualifying Project, Worcester Polytechnic Institute. https://wp.wpi.edu/puertorico/projects/2023-jan-mar/facilitating-a-network-of-rescued-schools-in-puerto-rico/

Ambrad, N., Devisetti, V., Sequira, S., & Troiano, M. (2025). *Growing Together: Strengthening la Conde's Community Kitchen & Programs*. Interactive Qualifying Project, Worcester Polytechnic Institute. https://wp.wpi.edu/puertorico/projects/2025-jan-mar/la-conde/

Blanchard, A., Phillips, G., Logan, J., Reynolds, M., & Martins, N. (2024). *Citizen Science Noise Monitoring for the Calle Loíza Community.* Interactive Qualifying Project, Worcester Polytechnic Institute. https://wp.wpi.edu/puertorico/projects/2024-jan-mar/wip-4/

Brennan, J., Mamayek, T., Nearing, I., & Strickland, A. (2021). *Exploring Water and Power Resiliency for the Cubuy Community Center in Canóvanas, Puerto Rico.* Interactive Qualifying Project, Worcester Polytechnic Institute. https://wp.wpi.edu/puertorico/projects/2021-fall/cubuy-roofs/

Checko, T., Goel, A., Sumner, A., & Wood, N. (2024). *Designing a Resilience Hub for Vulnerable Populations in Las Carolinas*. Interactive Qualifying Project, Worcester Polytechnic Institute. https://wp.wpi.edu/puertorico/projects/2024-jan-mar/wip1/

Chesanek, R., Kerekon, A., Lassard, H., & Russell, E. (2023). *Strengthening La Goyco's Volunteer Program.* Interactive Qualifying Project, Worcester Polytechnic Institute. https://wp.wpi.edu/puertorico/projects/2023-mar-may-2/goyco/

Cohen, J. R. (2001). Abandoned Housing: Exploring Lessons from Baltimore. *Housing Policy Debate, 12*(3), 415–448. https://doi.org/10.1080/10511482.2001.9521413

Conway, W., Delo, M., Tweedie, S., & Willey, N. (2021). *Advancing the Cubuy-Lomas Community Services and Development Center in Puerto Rico.* Interactive Qualifying Project, Worcester Polytechnic Institute. https://wp.wpi.edu/puertorico/projects/2021-fall/cubuy-lomas-center/

Cranston, S. (2017). *School Closures & Community Hubs: Examining Livability in Ontario Through School Closures and the Community Hub Framework* (Master's Thesis). Queen's University.

Cull, P., (2019). *Community-Based Disaster Response Teams for Vulnerable Groups and Developing Nations: Implementation, Training, and Sustainability* (Master's Thesis). Massey University, Wellington, New Zealand.

Cutter, S. L., Barnes, L., Berry, M., Burton, C., Evans, E., Tate, E., & Webb, J. (2008). A Place-Based Model for Understanding Community Resilience to Natural Disasters. *Global Environmental Change, 18*(4), 598–606.

Davis, J. M., Checko, T., Sumner, A., & Molinari, S. (2025). The Disaster Response Mobilization System: A Community-Based Disaster Preparedness Strategy in Puerto Rico. *Disasters, 49*(4), Article e70012.

Davis, J. M., Reyes, M., Abrogar, J., Bourgoin, J., Brown, M., Kellum, E., Polito, F., & Jiusto, S. (2023). Puerto Rico's Rescued Schools: A Grassroots Adaptive

Reuse Movement for Abandoned School Buildings. *Social Sciences, 12*(12), 662–677.

Diaz, F., Johnson, A., Quattlander, M., & Turner, B. (2024). *Las Carolinas Community Hub: Unifying Progress with El Centro de Apoyo Mutuo*. Interactive Qualifying Project, Worcester Polytechnic Institute.

Duggan, M., Ng, T., Siepmann, R., & Benchikhi, Y. (2024). *Project Salvage: Enhancing an Innovative Clothing Donation Program in Cubuy-Lomas, Puerto Rico*. Interactive Qualifying Project, Worcester Polytechnic Institute. https://wp.wpi.edu/puertorico/projects/2024-jan-mar/wip-5/

Fallon, W., Herzig, N., Lufkin, A., & O'Mullan, B. (2023). *Advancing Community Climate Adaptation: Case Study on Resilience Centers in Puerto Rico*. Interactive Qualifying Project, Worcester Polytechnic Institute. https://wp.wpi.edu/puertorico/projects/2023-mar-may-2/ccan/

Gagnon, M., Howland, J., Little, M., & Oliveira, C. (2025). *Developing Program Efficiency for a Mutual Aid Organization in Puerto Rico Using Accessible and Sustainable Systems*. Interactive Qualifying Project, Worcester Polytechnic Institute. https://wp.wpi.edu/puertorico/projects/2025-jan-mar/cam-las-carolinas/

Govani, S., Hadad, B., MacDonald, S., & Pina, J. (2023). *The Changing Face of La Goyco: Examining the Impact of Gentrification on a Historic Community*. Interactive Qualifying Project, Worcester Polytechnic Institute. https://wp.wpi.edu/puertorico/projects/2023-jan-mar/the-changing-face-of-la-goyco-examining-the-impact-of-gentrification-on-a-historic-community/

Hasbrouck, M., Marcinkowski, J., Brown, J., & Schletzbaum, K. (2024). *Strengthening Emergency Preparation & Management in the La Goyco Community*. Interactive Qualifying Project, Worcester Polytechnic Institute. https://wp.wpi.edu/puertorico/projects/mar-apr-2024/goyco/

Hinojosa, J., Meléndez, E., & Severino Pietri, K. (2019). *Population Decline and School Closure in Puerto Rico*. Center for Puerto Rican Studies (CENTRO).

Holmes, A., Mehta, N., Perriello, J., & Wertz, E. (2023). *Developing Resilience Strategies in Cubuy-Lomas, Puerto Rico*. Interactive Qualifying Project, Worcester Polytechnic Institute. https://wp.wpi.edu/puertorico/projects/2023-jan-mar/developing-resilience-strategies-in-cubuy-lomas-puerto-rico/

Jiusto, S., McCauley, S., & Stephens, J. (2013). Integrating Shared Action Learning into Higher Education for Sustainability. *The Journal of Sustainability Education, 5*, 1–22.

Kawatkar, P., Morris, N., Perry, M., & Potter, J. (2024). *Envisioning La Conde: Empowering Communities Through Technology and Emergency Plans*. Interactive Qualifying Project, Worcester Polytechnic Institute.

Li, E. Indipiginja, J., Smith, C., & Sawyers, T. (2025). *Strengthening Emergency Management with the La Goyco Community*. Interactive Qualifying

Project, Worcester Polytechnic Institute. https://wp.wpi.edu/puertorico/projects/2025-mar-may/goyco/
Martínez M., & Lee, J. (2018). *Fighting Manufactured Chaos in Puerto Rico's Public Schools. Rethinking Schools.* https://rethinkingschools.org/articles/fighting-manufactured-chaos-in-puerto-rico-s-public-schools/ (accessed 28 March 2025).
Milman, O. (2017, November 9). Six Weeks After Hurricane María, Puerto Ricans Still Waiting for Help from FEMA. *The Guardian.* https://www.theguardian.com/world/2017/nov/09/six-weeks-after-hurricane-maria-puerto-ricans-still-waiting-for-help-from-fema (accessed 2 April 2024)
Montalto, F., & Sheller, M. (2025, June 4). *Strengthening Social Infrastructure for Climate Resilience.* The Nature of Cities. https://www.thenatureofcities.com/TNOC/2025/06/04/strengthening-social-infrastructure-for-climate-resilience/
People for Education. (2009). *School Closings and Declining Enrolment in Ontario.* http://www.peopleforeducation.ca/wp-content/uploads/2011/07/Annual-Report-on-Ontario-Schools-2009.pdf
Rivera Rivera, E. R. (2022). *Schools of Encanto: Puerto Rico's Schools Closure Policy* (Master's Thesis). John Hopkins University, Baltimore, MD, USA.
Rodríguez, J. (2015). *Walkability Study for School Accessibility: Case Study of the San Juan, Puerto Rico Elementary Schools* (Master's Thesis). University of Southern California, Los Angeles, CA, US.
Roque, A., Pijawka, D., & Wutich, A. (2020). The Role of Social Capital in Resiliency: Disaster Recovery in Puerto Rico. *Risk, Hazards & Crisis in Public Policy, 11*(2), 204–235.
Rosenbaum, M., Kim, K. K., Ramirez, G. C., Orejuela, A. R., & Park, J. (2019). Improving Well-Being Via Adaptive Reuse: Transformative Repurposed Service Organizations. *The Service Industries Journal, 41*(3–4), 223–247. https://doi.org/10.1080/02642069.2019.1615897
Strier, R. (2014). Fields of Paradox: University–Community Partnerships. *Higher Education, 68*, 155–165.
U.S. Census Bureau. (n.d.). *Annual Estimates of the Resident Population for Selected Age Groups by Sex for the United States, States, Counties, and Puerto Rico Commonwealth and Municipios: 2001 to 2021.* Retrieved March 28, 2025, from https://www.census.gov/data/tables/time-series/demo/popest/2020s-detail-puerto-rico.html
U.S. Census Bureau. (2021). *Explore Census Data.* Retrieved March 28, 2025, from https://data.census.gov/profile/Lomas_barrio,_Can%C3%B3vanas_Municipio,_Puerto_Rico?g=060XX00US7202946669
Weerts, D. J. (2005). Facilitating Knowledge Flow in Community-University Partnerships. *Journal of Higher Education Outreach and Engagement, 10*(3), 23–38.

Witten, K., McCreaner, T., Keams, R., & Ramasubramanian, L. (2001). The Impact of School Closure on Neighbourhood Cohesion: Narratives from Invercargill, New Zealand. *Health & Place, 7*, 307–317. https://doi.org/10.1016/S1353-8292(01)00023-5

Yedidia, D. R., Moore, E., Rivera, L. G., & Arabía, M. M. (2020). *Puerto Rico's Public School Closures: Community Effects and Future Paths*. Othering & Belonging Institute.

CHAPTER 4

Community Climate Adaptation in the U.S. Virgin Islands

Sol Giesso, Jonathan Chang, and Sarah Strauss

Abstract This chapter reflects on the methods and findings of a place-based research project conducted as a collaboration between Community Climate Adaptation (CCA) Master's students and faculty at Worcester Polytechnic Institute and the University of the Virgin Islands under the Caribbean Collaborative Action Network (CCAN), a NOAA climate adaptation partnership. The goal of the project was to develop a baseline understanding about climate change threats and action in the USVI.

S. Giesso (✉) · J. Chang
Community Climate Adaptation, Worcester Polytechnic Institute, Worcester, MA, USA
e-mail: sgiesso@wpi.edu

J. Chang
e-mail: jchang1@wpi.edu

S. Strauss
Department of Integrative and Global Studies, Worcester Polytechnic Institute, Worcester, MA, USA
e-mail: sstrauss@wpi.edu

S. Molinari and M. Sheller (eds.), *Building Equitable Climate Adaptation Partnerships in the US Caribbean*,
https://doi.org/10.1007/978-3-032-15727-0_4

Our study characterized the USVI community's perception of climate risks, vulnerabilities, and resilience while assessing the alignment—or lack thereof—among these perspectives and existing policy frameworks. Through a combination of participant observation, ethnographic interviews, document and policy review, and gap analyses, this study captures the voices of residents, policymakers, scientists, and community leaders across St. Thomas, St. John, and St. Croix, considering how these diverse stakeholders understand and navigate climate change. We explore the community-engaged and participatory research approach that defined and guided this project, from initial brainstorming to the ultimate recommendations of the study. Key themes include the impacts of climate change on infrastructure, food and water security, health, and the economy, as well as challenges faced by external researchers in making meaningful connections with the USVI community.

Keywords Climate adaptation · Community engagement · U.S. Virgin Islands · Participatory research · Climate change governance

1 Introduction

In this chapter, we present some ways that the training of graduate students in the emergent field of community climate adaptation has helped to develop a strongly community-engaged and decolonial research approach to define and guide part of our NOAA Caribbean Collaborative Action Network (CCAN) work in the U.S. Virgin Islands (USVI). CCAN's leadership is based in Puerto Rico, with collaborators across the mainland U.S. and the USVI, but there are still limitations in capacity (including linguistic challenges and overall human/other resource availability) to address specific research and action needs in the USVI. For this reason, co-authors Sol Giesso and Jonathan Chang, a team of Master's students in WPI's graduate program in Community Climate Adaptation (CCA)—supervised by Professors Sarah Strauss and Seth Tuler—have provided additional research support to help meet CCAN goals. CCA is a 3-semester, fieldwork-based program that builds on WPI's distinctive interdisciplinary approach to project-based learning. Designed to give students field-based training to work collaboratively within communities and organizations, it is a joint program between the Department of

Integrative and Global Studies (DIGS) and WPI's Department of Civil, Environmental, and Architectural Engineering.

The program is designed with theoretical and methodological breadth in mind, taking an explicitly justice-centered approach to local climate adaptation. Students take courses that characterize how climate change impacts food, energy, water, the built environment, and social-ecological systems; they learn the basics of environmental planning and qualitative methods, and gain opportunities to analyze and compare local contexts for adapting to climate change. Specialized electives in GIS and more technical or administrative skills are also offered. The program is centered on collaboration, both in terms of the student teams who produce a joint thesis project and with respect to the students' development of community engagement processes and ethical practices for sustainable change. CCA's team-based and community-centered research approach resonates strongly with the work of Goitom and Hepburn (2025) on "duoethnography" as a decolonial pedagogical methodology. It extends WPI's project-based learning approach, a pedagogical technique that aspires to justice, diversity, and equity in the classroom and beyond (cf. Pfeifer & Stoddard, 2019), established in 1970 as a way of transforming engineering education.

This project was conducted by a team of two students, co-authors Giesso and Chang, whose own lived experiences in Samoa (which was colonized by Germany and recognized as the first Pacific Island to attain political independence in 1962) and Argentina brought shared postcolonial understandings to their research. They, in turn, were guided in their research by diverse local partners employed at the University of the Virgin Islands, an HBCU (historically Black college or university). The multiple interpretive frames that the students and their interlocutors, from mentors to discussants in the field, brought to this project reflect the kind of value that plural (duo and beyond) ethnography can bring to a project: "duoethnography emphasizes the role of emotional and bodily knowledge that is holistically amassed through the research process. In this way, those embarking on the inquiry can use their own biographies as sites of research, fostering the creation of dialogic narrative" (Goitom & Hepburn, 2025, p. 9).

The USVI are located in the Caribbean Sea approximately 1100 miles southeast of Miami and 40 miles east of Puerto Rico. The territory comprises four principal islands—St. Croix, St. John, St. Thomas, and Water Island—as well as several small islands and cays adjacent to Puerto Rico and the British Virgin Islands, with varied topography including mountainous interiors, low-lying coastal plains, and over 175 miles of coastline. A U.S. National Park covers most of St. John, and the territorial governance structure manages the four main islands as just two districts, with the majority of people residing on St. Thomas (including administrative leadership for St. John/Water Island) and St. Croix. Indigenous groups—including the Arawak, Taino, and Carib—inhabited the Caribbean long before the arrival of Christopher Columbus in 1493; however, European colonization, warfare, and epidemic diseases led to the near decimation of these communities by the late sixteenth century (Fernandez, 2022). Successive colonial regimes by Spain, the Netherlands, and Denmark claimed and reconfigured the islands, which became the Danish West Indies in the seventeenth century. Denmark maintained a plantation economy based primarily on enslaved African labor until emancipation in 1848. Denmark then sold the territory to the United States in 1917 (U.S. Department of State, 2009), and it remains a territory without political representation.

Despite holding U.S. citizenship since 1927, Virgin Islanders lack full political rights. The territory elects a non-voting delegate to the U.S. House of Representatives, who may serve on committees and participate in debate but cannot cast floor votes. Further, residents cannot vote in the U.S. presidential elections, receiving only symbolic participation through party primaries (Wang, 2025). These democratic limitations are often described as a form of "taxation without representation," since Virgin Islanders are subject to federal law and policy decisions without having a direct voice in them. This condition reflects the coloniality of territorial governance shared between Puerto Rico and the USVI: Virgin Islanders are integrated into the U.S. system yet excluded from its most basic democratic rights, reinforcing their status as second-class citizens in a politically marginal but strategically important territory (Wang, 2025). In such contexts, research must move beyond extractive methods toward decolonial approaches that elevate community voices, recognize historical asymmetries, and co-produce knowledge in ways that resist reproducing colonial hierarchies.

The current population of the USVI is 84,000, down from 108,000 in 2000 (United States Virgin Islands Population 2025, 2025). The official language is English, but about 20% are Spanish speaking, mostly natives of Puerto Rico and the Dominican Republic. Many residents are descendants of enslaved Africans, which has shaped the Islands' cultural and social identity; well over 70% of residents identify as Black or Afro-Caribbean (Federal Reserve Bank of New York, 2025). The Islands have also seen migration of Afro-Caribbean people from other islands such as St. Kitts and Nevis, Dominica, and Antigua and Barbuda in recent years, adding to the territory's demographic complexity. Socio-economic challenges such as high unemployment and income inequality, as well as the general precarity of the energy and food systems and a limited transportation sector, all influence the Islands' ability to respond to and recover from climate impacts. The heavy dependence on tourism and reliance on imports for more than 95% of their food makes the situation in the USVI particularly challenging in the face of climate change. Like many island territories, the USVI is experiencing a demographic shift, with an aging population and limited opportunities for young people. This transition places a growing demand on healthcare, social services, and infrastructure, and presents challenges for workforce replacement. The islands have been experiencing extreme and continuing impacts from back-to-back category 5 Hurricanes Irma and Maria in 2017.

By considering the challenges we faced as a collaborative team of external and locally based researchers in approaching local populations, we suggest ways to navigate those barriers to make meaningful connections with the USVI community. We highlight the importance of working with a local liaison who can support community engagement by helping to identify and connect with hard-to-reach—or simply reluctant—research participants, develop a sense of local confidence in the study, and form stronger bonds across collaborating organizations. We also consider cross-cutting community concerns regarding governance for climate change adaptation. Our discussants identified six critical areas: the need for local solutions rooted in the USVI's unique context, the development and funding of policy for climate change resilience, policy implementation limited by workforce capacity, the lack of climate change leadership, the importance of education to foster awareness and action, and the need for more collaboration with other countries and regions. We then present recommendations that emphasize how the integrated nature of climate resilience in the USVI requires cohesive efforts spanning policy reform, community engagement, and innovative partnerships that transform the USVI's vulnerabilities into opportunities for a more sustainable and equitable future. To begin, however, we want to focus on our methodological approach to site-based collaboration in the context of the USVI, with an emphasis on community engagement.

2 Methodological Approach and Community Engagement

This project explored how communities in the USVI understand climate risks, vulnerabilities, and adaptation, and how these perspectives align (or do not) with existing policies. Fieldwork in the USVI took place from January to May of 2024, and involved participant observation as well as 41 semi-structured interviews (Bernard, 2017) with a diverse group of participants, including residents, scientists, policymakers, and nonprofit leaders across the islands of St. Thomas, St. Croix, and St. John.

We analyzed the coded data by participant role, island, and topic to uncover commonalities and differences across the USVI archipelago. In addition to the interview data, we conducted a policy review of four key policy documents: the Agricultural Plan,[1] the Hazard Mitigation and Resilience Plan (HMRP),[2] Vision 2040,[3] and the Comprehensive Land and Water Use Plan (CLWUP).[4] This thematic mapping helped us identify areas of convergence and divergence between community concerns and policy.

2.1 *Importance of a Local Liaison and Informal Engagement*

A critical element of our methodology was working closely with a local community engagement specialist affiliated with the Virgin Islands Experimental Program to Stimulate Competitive Research (VI-EPSCoR) and the University of the Virgin Islands (UVI), as well as the Caribbean

[1] The Agricultural Plan outlines strategies to revitalize local farming, reduce food imports, and strengthen food security (U.S. Virgin Islands Department of Agriculture, 2021).

[2] The Hazard Mitigation and Resilience Plan is the territory's FEMA-approved disaster management framework, aimed at reducing vulnerability to hazards through risk assessment, mitigation, and recovery planning (Virgin Islands Territorial Emergency Management Agency, 2024).

[3] Vision 2040 is the USVI's economic development plan that incorporates sustainability and climate resilience into broader economic diversification goals (U.S. Virgin Islands Economic Development Authority, n.d.).

[4] The Comprehensive Land and Water Use Plan is a territory-wide planning framework that guides land use, zoning, and infrastructure development with attention to environmental protection and climate risk reduction (U.S. Virgin Islands Department of Planning and Natural Resources, 2024).

Collaborative Action Network (CCAN), and funded in part through another NOAA grant associated with the CCAN network.[5] Having a local liaison played a pivotal role in guiding the research in a way that was respectful, culturally attuned, and grounded in local realities. Her contributions went far beyond logistics. During the research process, her contributions drew on a deep understanding of the territory's history, geography, and social dynamics, which enabled more meaningful engagement with participants. She often knew the background behind community tensions or place-based references. This ability to contextualize responses enriched both the quality of the data and the level of comfort felt by the participants when, for example, they all shared moments of continued grief for the ongoing hurricane trauma.

In addition to her role, we also found that informal methods of engagement, typical of participant observation (Bernard, 2017), such as mutual introductions, attending local events, and initiating casual conversations, played a significant role in building trust. While official outreach channels generated interest, they often required extensive follow-up. In contrast, personal invitations and community referrals were more efficient and led to more enthusiastic participation. These informal "back door" approaches were not only more effective at establishing rapport, but also crucial for navigating research fatigue and skepticism, particularly with communities that have been repeatedly studied in the aftermath of disasters such as Hurricanes Irma and Maria, and more recently, COVID-19. As Francis et al. (2021) demonstrate, the post-disaster context in the USVI placed heavy demands on frontline workers and residents alike, with many experiencing burnout and exhaustion under the strain of both service provision and repeated research, reinforcing skepticism toward outside studies—being perceived as extractive or one-sided. Against this backdrop, our local liaison and informal community channels ensured that the study remained rooted in the everyday realities of life in the USVI.

[5] NOAA-BIL: *Improving Engagement Methods for Coastal Resilience and Reducing Climate Risk: Bridging Learning Networks From the Urban Northeast (CCRUN) to the US Caribbean (CCAN)*, (PI) Sheller, Mimi (Worcester Polytechnic Institute), Grant Award Number NA23OAR4310396.

2.2 *Local Confidence in the Study*

Upon arrival in the territory in January of 2024, the team was met with a fair degree of caution and hesitation. This skepticism was understandable: Communities in the USVI, like many other vulnerable regions, have often been researched by outsiders without seeing any tangible results or meaningful change. As noted by Coomansingh (2025), Caribbean populations face particular risks of research fatigue due to their small size and frequent targeting for studies, often leaving residents wary of further participation when they perceive little or no benefit. However, as we continued to engage with residents and community leaders by attending events, being transparent about our goals, and showing up consistently, participants began showing genuine interest in our work. Signs of growing confidence appeared through word-of-mouth: participants not only recommended others we should speak with, but also began telling others about our research and inviting us to attend local events. Their trust had extended beyond individual interviews. Perhaps most telling was the shift among our local UVI/CCAN partners who were not social scientists. While initially supportive yet cautious, they later came to see the potential for our qualitative study to provide insights and narratives that could complement quantitative research and inform future climate initiatives.

Giesso and Chang's personal identities also played a subtle yet important role: As a Spanish-speaking Argentinian and an islander from Samoa, these team members were often seen as outsiders who still shared aspects of linguistic and cultural familiarity. In Giesso's case, her Spanish was frequently linked by participants to Puerto Rico—the larger U.S. territory in the Caribbean with close historical, cultural, and migration ties to the USVI—creating a bridge of shared language and regional affinity despite her being from Argentina. For Chang, although he was from the independent state of Samoa, participants connected his background to American Samoa, a U.S. territory in the Pacific, which prompted conversations about shared structural realities of island territories under U.S. governance with limited political representation and dependency on external decision-making. These exchanges helped foster trust and connection and allowed participants to situate their own experiences within a broader global context.

2.3 *Stronger Bonds Among Collaborating Organizations*

Beyond our interactions with individual participants, this project contributed to stronger institutional relationships across the territory. It reaffirmed and deepened the existing collaboration between WPI and UVI, providing a shared experience grounded in fieldwork and co-produced knowledge. More than just a student project, the study became a vehicle for strengthening communication and trust between researchers on both sides of the partnership.

It also supported broader collaboration across sectors. Through this work, we built connections with various USVI agencies, community-based organizations, and nonprofit groups. Many of these relationships were developed informally, but they led to meaningful conversations about policy, planning, and future partnerships. We engaged with CCAN students and faculty on a regular basis, which helped frame our study within a regional context and opened doors to cross-island learning and support. These bonds have laid a strong foundation for future research, capacity building, and adaptation efforts centered on community voices and inclusive planning.

3 Findings and Discussion

In this section, we present our findings on the impacts of climate change on infrastructure, society, culture, and the economy of the USVI. We end with a discussion of how climate adaptation governance might evolve in ways that bridge policy gaps and address community concerns.

3.1 *Infrastructure Bottlenecks: Ports of Entry and Roads*

Climate change affects the USVI's ports of entry and road infrastructure, both of which are critical to essential goods and services, emergency response, and the economy. They face increasing vulnerability from hurricanes, floods, erosion, and sea level rise, which disrupt operations and threaten safety. During Hurricanes Irma and Maria, the Virgin Islands Port Authority (VIPA) facilities sustained $80–85 million in damage, severely affecting locals, according to a VIPA official.

Looking into the coming decades, a discussant expressed concerns, saying, "If you've got an economy that [is] dependent on your ports, what happens if you anticipate in 15 years the ports are going to be underwater?" With more frequent flooding, roads are also at risk. Inadequate drainage often renders roads impassable, isolating residents and delaying emergency response. Restoring access is further hampered by a shortage of skilled labor: "We have very, very limited skilled construction-based labor; we have very few large construction firms that have the capacity to do big jobs."

3.2 *Necessities at Risk: Water and Food*

Climate change has significant impacts on water and food in the USVI. The water infrastructure in the USVI faces ongoing challenges in meeting the population's water needs, as the Water and Power Authority (WAPA) supply does not reach all communities. The lack of access to municipal water sources forces residents to rely on rain-catchment cisterns, which are highly vulnerable to shifting rainfall patterns. One resident highlighted this vulnerability: "There is no WAPA… We rely on the rain. If there is no rain, we have no water." Another asked, "A cistern that has always been big enough…, [is it] still big enough?" In response, people dig more cisterns and wells, depleting the aquifers and further accelerating water shortages, explained an officer of the USVI Department of Planning and Natural Resources (DPNR).

Importantly, water scarcity and irrigation costs limit food production and strain farmers amid changing climate conditions. Without irrigation, many farmers have "lost livestock or had to cull their herds because they couldn't feed them," said an interviewee. Today, the USVI imports between 97 and 99% of their food, despite having had a strong agricultural history (DOA, 2021; VITEMA, 2024), something that two discussants described as "shameful" and "embarrassing." The loss of agriculture carries an emotional weight. Major development projects in the later half of the twentieth century—namely the St. Croix refinery—seized fertile farmland and extinguished the promise of homestead agriculture, leaving many Virgin Islanders feeling stripped not only of their livelihoods but of a cultural inheritance replaced with further colonial control (Bond, 2022). Thus, the further loss of agriculture today looms heavy on residents who risk losing traditional techniques and food-based cultural events (see Fig. 1).

Fishing was also an area of concern for locals, as the practice sustains local communities and is part of the Virgin Islands' cultural identity and

Fig. 1 Vendor at the Bordeaux Agricultural Fair (Photograph by S. Giesso, 2024)

tourist attraction. Warming seas and reef degradation threaten marine ecosystems. Because almost all of the food consumed in the USVI is imported, local resources like fisheries are especially important. As one restaurant owner put it, "You can't get fish, you can't sell!... The conch used to come in heaps... The specials like whelks and conch, they're special. So when Carnival[6] time comes... I don't know if we're going to have them anymore." Conservation approaches sometimes conflict with local fishing traditions. A resident expressed, "There's pressure on our culture because of the need to preserve coral fish, which is what we've

[6] The Carnival is an annual festival in many Caribbean islands created by enslaved and freed Africans. In the USVI, the practice dates back to the eighteenth century (Bastian, 2009). Today, it celebrates local food, music, and history and attracts thousands of attendees, both locally and from the diaspora (U.S. Virgin Islands Department of Tourism, 2025).

eaten for thousands of years… So how do we balance what our cultures have always done with the need for adapting to the changes on our reefs?" The answer to this question perhaps lies in the application of a new form of "decolonial ecology," in which the "critique of the destruction of the planet's ecosystems is then intimately tied to the critique of colonial and postcolonial dominations and to demands for equality" (Ferdinand, 2022, p. 14).

3.3 From Devastation to Adaptation: Strengthening Buildings and Homes

The devastation from the 2017 Hurricanes Irma and Maria left lasting marks on both private and public infrastructure. For one, the already limited housing stock suffered from severe damage, displacing many Virgin Islanders off-island due to a lack of affordable options. Building codes across islands are meant to help sustain buildings against storms, but as a government official at the DPNR explained, "Our building codes to meet storms should be implemented, and they should be subsidized too… But they're not always implemented." Traditional designs like hip roofs, known for wind resistance, are being reconsidered in rebuilding efforts (Fig. 2). One resident said, "Our ancestors figured it out because they had to live in the climate, as it was."

Reconstruction remains slow due to bureaucratic hurdles and fraud. For instance, under rebate programs, the Federal Emergency Management Agency (FEMA) provided wood lumber for rebuilding roofs and blue tarpaulins for temporary coverage. However, years later, many homes still only have blue "roofs" (Fig. 3) while the wood planks remain unused (Fig. 4). The unused lumber became a widespread controversy, culminating in a federal fraud indictment of the two government officials and one of their wives, who allegedly profited by storing the lumber rather than distributing it (Alred, 2024). Insurance claim delays also slowed rebuilding efforts. A homeowner expressed, "My insurance company went bankrupt… I was very fortunate that I… cashed that check before they went bankrupt. Some people weren't that fortunate." Home insurance costs—now the highest in the U.S.—have surged since 2017 (VITEMA, 2024), creating a burden for many in the territory.

Furthermore, damaged housing contributes to mental health struggles, an aspect of recovery emphasized less frequently. One resident remarked, "Just the reality of the fact that the hurricanes took place in 2017, and

Fig. 2 Traditional hip roofs that survived the hurricanes (Photograph by S. Giesso, 2024)

Fig. 3 Blue tarpaulin installed on a damaged roof in the aftermath of Hurricanes Irma and Maria, still there seven years later (Photograph by S. Giesso, 2024)

Fig. 4 Undistributed FEMA-aid lumber at a school in St. Croix (Photograph by S. Giesso, 2024)

we are in 2024… still seeing buildings that need to be at least removed so that people don't suffer from post-traumatic stress." Beyond building resilience in the literal sense, the literature suggests that humane disaster response should incorporate trauma-informed care (Kusmaul, 2020).

3.4 Navigating Reliability and Costs: Energy Access and Diversification

As an isolated power grid, island-wide outages are a constant challenge. A resident explained, "The power situation is really dire… We go out of power at least once a month, usually closer to once a week, for hours at a time." After Hurricanes Irma and Maria, power restoration took months. Locals emphasized the importance of having their own generators for situations like these, "You have to have alternative power." However, generators are costly, environmentally damaging, and are just a "quick fix" to a deeper issue. One discussant expressed their disappointment with the frequent outages by saying, "We don't live in a third-world country, this is America." This remark highlights the colonial tension between expectations of modern U.S. American services and the lived experience of chronic outages in the marginal USVI, a stark and ironic contrast that residents often observed in this and other areas.

When power is available, electricity costs in the USVI are the most expensive in the United States, and among the highest globally, with residents paying up to 43 cents per kilowatt-hour, almost three times the national average. These exorbitant costs disproportionately affect low-income families and seniors, many of whom cannot afford basic utilities. One discussant explained, "The outrageous cost of energy and power here on the islands—half of them can't afford AC if they need it, can't afford lights if they need it, and even refrigeration at times." Businesses also bear the brunt of high electricity costs. For many, energy expenses represent a significant portion of their operating budgets, forcing them to lower wages or delay investments in growth, as one business owner expressed.

Residents advocated for renewable energy sources that would increase resilience. For example, self-reliance on renewables reduced power interruptions during the hurricane recovery period. A nonprofit leader recalled, "We were able to solarize a good portion of our headquarters... And as a result of that foresight years ago, years before Maria and Irma, we had electricity, internet and phone within four days of the hurricane." Still, access to renewables remains inequitable. As one resident expressed, "What we're seeing in the territory right now is that everyone who is going off grid is the upper end of the income spectrum. So they are leaving the cost burden on the lower half of the spectrum." In other words, these decentralized efforts may come at the expense of more centralized public infrastructure provisions that serve all residents, raising important questions for ethical and equitable climate adaptation.

3.5 The Fragility of "Paradise": Climate Risks to the Tourism Economy

The degradation of the natural environment in the USVI raises significant concerns among residents, as the territory's white sand beaches, turquoise waters, and rich marine biodiversity (see Fig. 5), which draw tourism and define local culture, are increasingly under threat. As one resident put it, "We sell beach to our tourists. We don't sell them anything else because we've got nothing else to sell them." Another resident reflected, "The more I think about it while we go through this interview... The basic question of, on what will the economy of the Virgin Islands [depend]?... It requires a potential redefinition of the entire Virgin Islands... That would probably be the primary reason why we haven't had as much serious thought and action as the situation requires. It's probably a little bit too terrifying." Residents' concerns resonate with broader Caribbean

Fig. 5 View of Trunk Bay in St. John (Photograph by S. Giesso, 2024)

debates about climate, tourism, and local agency. Sheller (2020) emphasizes that islands across the region face similar pressures, where ecological vulnerability intersects with economic dependence on tourism.

The main historical—and thus touristic—districts, Frederiksted, Christiansted, and Charlotte Amalie, are located in vulnerable coastal areas. A government worker said, "[Due to] rising water tables, I would say that the viability of the three largest towns Charlotte Amalie, Christiansted, and Frederiksted, are all at serious risk." Another discussant added, "The major restaurants, shops, people's businesses, are along that coastline." Both airports also lie in flood- and erosion-prone areas. It is important to note that while this section focuses on tourism, the factors that attract tourists to the USVI are deeply intertwined with the "culture, pride, history, and quality of life of [Virgin Islanders]." The loss of beaches and historical districts not only impacts the tourism industry but also profoundly affects the local community, as these features are central to locals' identity and way of life.

Additionally, extreme climate events can affect the ability of planes and ships to reach the USVI. The hurricane season in the USVI dictates tourism seasons, so if the boundaries of the seasons expand, the tourism season may shorten. These issues leave residents feeling powerless: with a small carbon footprint and no ability to vote in presidential elections, there is little the USVI can do to mitigate carbon pollution. As one resident put it, "Unfortunately, we get thrown into a situation where we're reacting to situations that really weren't created by these smaller spaces. And we are vulnerable, because... our natural environment, it's what we depend on." This vulnerability, however, extends beyond environmental dependence. Scholars such as Sultana (2022) have described the "coloniality of climate," highlighting how histories of colonialism and neocolonial economic structures have left island communities politically, socially, and economically exposed to external forces. In the USVI, these historical legacies compound ecological vulnerability, shaping how residents experience and respond to climate-related disruptions.

3.6 Governance for Climate Change

In analyzing governance for climate change, our interviews and policy review revealed six critical cross-cutting themes. They include the need for local solutions rooted in the USVI's unique context, the development and funding of policy for climate change resilience, policy implementation limited by workforce capacity, the need for more regional collaboration, the lack of climate change leadership, and the importance of education to foster awareness and action.

A cross-cutting theme throughout residents' concerns was the need for local solutions that value their traditional knowledge and lived experiences. As one resident put it, "I think we keep looking in the wrong direction for models. We keep looking north. And I think some of our greatest models will be due east or south, in developing countries." Another one added, "We're frontline because we're an island. And we're overlooked because we're an island. And we're overlooked because we're predominantly Black and Brown. And we're US, but not really." Their words point to climate coloniality embedded in political and economic systems: knowledge from marginalized communities is not only dismissed but actively silenced because of entrenched histories of racism and colonial domination (Sultana, 2022). For example, residents—especially farmers and fishers—felt that their input had been neglected in policy writing, weakening trust in institutions, and missing key opportunities.

This echoes broader critiques that adaptation planning often reproduces inequalities by privileging Global North models while overlooking locally grounded solutions (Andreucci & Zografos, 2022). Instead, studies show that the integration of traditional knowledge and modern science achieves more effective and sustainable implementation of climate change adaptation strategies (Kettle et al., 2014; Makondo &

Thomas, 2018). Taken together, these insights emphasize the urgency of community-based planning in the USVI and across the Caribbean, not only to imagine alternative futures but also to confront the structural conditions that perpetuate exclusion. Community-based research and planning efforts, including CCAN, can facilitate cross-island learning and the co-creation of adaptive strategies that reflect local priorities.

Residents hoped to see more collaboration within the Caribbean region or with other islands as a way to increase capacity and learn from one another. A historical legacy of fragmented nationalism in the Caribbean—that is, the fact that Caribbean societies were created under competing European empires, each with different languages, legal systems, religions, and political institutions—produces sharp divisions even among neighboring islands, complicating collaboration (Knight, 2012). Not only that, the USVI's territorial and isolated status often interfered with the ability of the islands to feel represented in federal priorities. One resident said, "Being on the national and international stage is going to require the region being cohesive. And because we're U.S., we don't participate in a lot of the regional work that's happening… CARICOM is doing work around climate, and they have a voice at the United Nations. We don't. We barely have a voice in the U.S. Congress." A more comprehensive climate adaptation strategy in the USVI could incorporate these collaborations as they build momentum for the Caribbean region and capacity for the USVI.

Residents also emphasized that adaptation requires not only good policy, but also funding and follow-through. Our review found that community concerns were addressed in documents like the Agricultural Plan, HMRP, CLWUP, and Vision 2040. However, the policies only go as far as their implementation does. At the time of writing, the Agricultural Plan has only received $500,000 of the requested $13,936,100 from the local government (Dias, 2024), and federal funding has also been lower than anticipated. The HMRP and CLWUP have not been funded, though the passing of the CLWUP in early 2025 was seen as a major step forward after decades of deadlock. As one resident said, "[The CLWUP is] the most climate resilient policy that the territory will have seen… So I think that matters greatly. And I think that it's achievable. So that's exciting."

Furthermore, even when funded, climate policies often stall between intention and action. This is evident in the case of FEMA recovery funding, where roughly $23 billion have been allocated to the U.S.

Virgin Islands but only about 24% has been expended, meaning 76% of available funds remain unused (Acosta et al., 2025; USVI Office of Disaster Recovery, 2025). Concurrently, Trump-era federal funding cuts have created a state of uncertainty of whether remaining funds will be de-obligated (Buchanan, 2025). These delays are largely caused by limited institutional capacity and an insufficient workforce, as a local economist we interviewed explained. Residents felt that the USVI should better leverage its existing resources to build capacity. As one resident put it, "The Virgin Islands is one of those rare places under the U.S. flag that does not have a municipal government. And so our nonprofits are not typical nonprofits. There are a number of nonprofits that are filling roles that in other, healthier communities, localized government would be responsible for." Stronger and more organized partnerships with these organizations could improve implementation, financing, and accountability.

These findings also highlighted the relevance and limits of our own training. Much of the CCA coursework emphasized urban vs. rural governance contexts or compared city, state, regional, and national-level frameworks for adaptation planning. Yet in the USVI, the absence of municipal government, the layered structure of territorial governance, and the blurred boundaries between nonprofit and public roles presented a unique landscape. Furthermore, our fieldwork revealed how these less typical government structures and embedded colonial relationships can complicate adaptation efforts, reinforcing the need for adaptive strategies that reflect realities on the ground. In fact, most of the climate adaptation initiatives mentioned by residents were led by nonprofit organizations such as Love City Strong, the St. Croix Community Foundation, the Coral Bay Community Council, and the Environmental Association of St. Thomas, who noticed and acted on the needs of their respective communities.

Beyond this, residents felt that there was not one entity currently leading climate change efforts in the territory. When asked about whether they could identify a key climate change figure in the USVI, one resident said, "I think that we have some [leaders] who may emerge in the coming months, but at current, no." Residents reminisced about the failure of a 2015 executive order that sought to build leadership around climate change in government through a USVI Climate Change Council. A community leader reflected, "I feel like sometimes we create these structures that don't actually have bite or power to do anything, and then they are just these paper councils. But it does seem like there's been a missing link with having somebody in Government House who's the champion for these issues." A government official added, "We do not need a fleet of climate change resilience experts in the government, you just need the right person. And that person needs to have the right resources and attention, bandwidth from others... And I think from there, a lot will change quickly."

Residents also believed that education is vital to addressing climate change in the USVI. Interview discussants felt that Virgin Islanders understood some topics, such as hurricane preparedness, very well. However, the connection between climate change and daily life was often unclear. One resident observed, "I don't really think it registers with the average citizen, how bad the situation is, what the future threats are." Another discussant emphasized the importance of climate education in K-12, "So when they become the leaders of tomorrow, [climate change] isn't something that they're just learning... It'll be something that they will be able to see that it's happening around them." Participants also felt that local government officials lacked climate literacy, and without an informed public demanding adaptation, progress stalled. One discussant explained, "This lack of awareness of climate change... translates into lack of political leadership... There's no accountability; there's no one demanding that my politician must have an environmental agenda." As a result, initiatives to build climate resilience and protect environmental health have failed to garner enough support and advocacy, perhaps failing to be framed and regarded as real concerns for the viability of the USVI.

4 Concluding Thoughts and Recommendations

This project provided great insight into climate change concerns and adaptation in the USVI. Findings from this study reveal that climate adaptation in the USVI is highly localized, emphasizing the importance of community-driven, context-specific responses. At the same time, our process elucidated ethical practices of ethnographic researchers working in foreign and delicate contexts. Below we provide some additional research recommendations relevant to other Caribbean climate adaptation contexts.

First, we encourage researchers to take a more decolonial approach, such as defining problems *with* the community, not *about* the community. Researchers studying the USVI should adopt participatory approaches that involve the community not only in data collection but also in defining objectives and methodologies. This approach requires researchers to shift their focus from pursuing questions that primarily serve their own academic or professional interests to addressing the pressing needs of the communities they study. Because there is significant skepticism of researchers, prioritizing transparency about reach and limitations is also important as it sets expectations for outcomes and follow-through. Through these deliberate actions, researchers can gain a degree of trust in the community, one that must be sustained with continued respect and empathy. Ultimately, this collaborative approach can help to ensure cultural relevance and empower communities as equal partners, leading to more impactful and meaningful research outcomes.

Secondly, in doing this, it is important to consider the specific context and local geographies of the community that may lead to spatial marginalization. In the USVI, for example, there is a looming sentiment that the island of St. Thomas is often prioritized in research, while St. Croix and St. John are cast aside. Thus, we made sure to spend enough time on each island and traveled frequently to meet interviewees and attend key community events, such as the St. Croix AgriFest.[7] Additionally, within each island, natural barriers (like hills) and infrastructure (like distance from a port) isolate certain communities, who have particular experiences regarding climate change and adaptation. When defining who or what is "local," it is important to consider multilayered spatial aspects.

Finally, we recommend ethical engagement as a means of learning from each other and fostering long-term partnerships with local partners. Working with a community liaison was of great value to our study since discussants felt more comfortable participating knowing that locals were involved and present during interviews. Even now, as our direct involvement in the project has moved to the next phase, local partners and engaged residents can continue to implement our findings and ask further questions. Additionally, engaging with community leaders expanded the reach of our results. For example, before Giesso and Chang departed the Islands in May 2024, they presented their preliminary findings to the

[7] The St. Croix Agrifest is an annual, three-day event held on the island of St. Croix, where dozens of growers and vendors showcase and sell fruits and vegetables, local dishes, jewelry, art, and clothing. Entertainment includes quelbe music, poetry, moko jumbies, quadrille, reggae music, fashion show, calypso, and gospel. The event attracts people from the entire region, averaging 30,000 attendees yearly (Stokes, 2023).

USVI Congress, ensuring that they received key information in an interactive manner rather than exclusively as an extensive written report months later.

This study highlights the pressing need for the U.S. Virgin Islands to address climate change with tailored, community-centric strategies that reflect the unique context of the territory. From fragile infrastructure and limited food sovereignty to underfunded policies and governance challenges, the vulnerabilities of the USVI illustrate the intricate links between climate impacts, social resilience, and policy effectiveness. Ultimately, we call for bold, integrative approaches to climate adaptation that acknowledge how climate resilience in the USVI cannot be achieved in isolation but requires cohesive efforts spanning policy reform, community engagement, and innovative partnerships. To that end, we also recommend investing in workshops and training opportunities—for both local professionals and visiting researchers—that can foster cross-sector collaboration, build technical capacity, and better connect island communities to broader regional and global climate adaptation efforts. By addressing the gaps identified in this study and leveraging the strength of its people and resources, the USVI can transform its vulnerabilities into opportunities for a more sustainable and equitable future.

Acknowledgements We would like to express our heartfelt gratitude to our local partners in the U.S. Virgin Islands, whose expertise and dedication were integral to this project. Special thanks to Dr. Gregory Guannel, Dr. Kim Waddell, Jeannine Francis-Brown, and Janet Turnbull-Krigger for their trusted input, thoughtful advice, and generous welcoming into their lives and communities. We also thank all those who contributed their careful assessment and intimate reflections of climate change impacts in the U.S. Virgin Islands, shaping the findings of this research and illuminating pathways for change. Thanks are due as well to Prof. Seth Tuler, WPI, for thoughtful insights and support as co-advisor for this project.

Competing Interests (1) This study was funded by NOAA BIL Grant #NA23OAR4310396-T1-01, *Improving Engagement Methods for Coastal Resilience and Reducing Climate Risk: Bridging Learning Networks From the Urban Northeast (CCRUN) to the US Caribbean (CCAN)*. PI, Mimi Sheller, WPI.

Ethics Approval Statement This study was performed in line with the principles of the Declaration of Helsinki. The Institutional Review Board at Worcester Polytechnic Institute (FWA #00030698; HHS # 00007374) reviewed this project and granted exempt status on January 26, 2024 (IRB-24-0397).

Informed consent to both participate in the research and publish the results was obtained from individual participants.

References

Acosta, J. D., Resnick, A. C., Kolb, C., Brown, L., Gill, D., Pint, E. M., Miller, C., Ali, R., Tierney, D., Locker, A. R., Yoho, K. D., Eckhause, J. M., & Lagerman, T. (2025). *Federal policy changes are needed to accelerate rebuilding infrastructure in the U.S. Virgin Islands*. https://www.rand.org/pubs/research_briefs/RBA3081-1.html#:~:text=This%20research%20brief%20outlines%20the%20findings%20and%20recommendations,gains%20from%20investments%20in%20rebuilding%20the%20physical%20infrastructure

Alred, T. (2024). *Wood rotting, hope fading: Seven years of unmet promises from EnVision Tomorrow*. The Virgin Islands Consortium. https://viconsortium.com/caribbean-exclusive/virgin-islands-wood-rotting--hopes-fading--seven-years-of-unmet-promises-from-envision-tomorrow

Andreucci, D., & Zografos, C. (2022). Between improvement and sacrifice: Othering and the (bio)political ecology of climate change. *Political Geography, 92*, Article 102512. https://doi.org/10.1016/j.polgeo.2021.102512

Bastian, J. A. (2009). 'Play mas': Carnival in the archives and the archives in carnival: Records and community identity in the US Virgin Islands. *Archival Science, 9*(1), 113. https://doi.org/10.1007/s10502-009-9101-6

Bernard, H. R. (2017). *Research methods in anthropology: Qualitative and quantitative approaches* (6th ed.). Rowman & Littlefield.

Bond, D. (2022). Anthropology in an age of upheaval: Reflections on environmental justice in the American empire of oil. *Antropologia Pubblica, 8*(1), 157–182.

Buchanan, D. (2025, February 20). *Senators ask, can Territory lose billions?* St. Thomas Source. https://stthomassource.com/content/2025/02/20/senators-ask-can-territory-lose-billions/

Coomansingh, K. (2025). *Research fatigue in small island countries*. Voices in Bioethics from the Caribbean Basin. https://www.thehastingscenter.org/wp-content/uploads/Kareem-Coomansingh-essay.pdf

Dias, D. (2024, May 4). *Local food and farm council provides update on how $500k will be distributed*. St. Thomas Source. http://stthomassource.com/content/2024/05/03/local-food-and-farm-council-provides-update-on-how-500k-will-be-distributed/

Federal Reserve Bank of New York, Research & Statistics Group. (2025). *U.S. Virgin Islands economic indicators*. Federal Reserve Bank of New York. https://www.newyorkfed.org/medialibrary/media/research/regional_economy/charts/Regional_VirginIslands

Ferdinand, M. (2022). *Decolonial ecology: Thinking from the Caribbean world* (A. P. Smith, Trans.). Polity Press.

Fernandez, N. (2022). *Chapter 12: The Caribbean: Introducing the region*. Gendered lives. https://milnepublishing.geneseo.edu/genderedlives/chapter/chapter-12-the-caribbean-introducing-the-region/

Francis, A. K., Clavier, N., & Hendrickson A. K. (2021). *Frontline government workers assessing post-disaster burnout and quality of life*. Natural Hazards Center. https://hazards.colorado.edu/public-health-disaster-research/frontline-government-workers-assessing-post-disaster-burnout-and-quality-of-life#:~:text=These%20groups%20endure%20additional%20stresses%20during%20a%20disaster,burnout%20and%20frontline%20workers%E2%80%99%20professional%20quality%20of%20life

Goitom, M., & Hepburn, S. (2025). Reframing metanarratives on Africa and the Caribbean through decolonial pedagogy: Duoethnography as embodied methodology. *Social Epistemology, 39*(5), 532–548. https://doi.org/10.1080/02691728.2025.2475461

Kettle, N. P., Dow, K., Tuler, S., Webler, T., Whitehead, J., & Miller, K. M. (2014). Integrating scientific and local knowledge to inform risk-based management approaches for climate adaptation. *Climate Risk Management, 4–5*, 17–31. https://doi.org/10.1016/j.crm.2014.07.001

Knight, F. W. (2012). *The Caribbean: The genesis of a fragmented nationalism* (3rd ed.). Oxford University Press.

Kusmaul, N. (2020). Role of trauma-informed care in disasters. *The Intersection of Trauma and Disaster Behavioral Health* (pp. 145–162). Springer International Publishing.

Makondo, C. C., & Thomas, D. S. G. (2018). Climate change adaptation: Linking Indigenous knowledge with Western science for effective adaptation. *Environmental Science and Policy, 88*, 83–91. https://doi.org/10.1016/j.envsci.2018.06.014

Pfeifer, G., & Stoddard, E. (2019). Equitable and effective student teams: Creating and managing team dynamics for equitable learning outcomes. In K. K. Wobbe, & E. A. Stoddard, Taylor & Francis Group (Eds.), P*roject-based learning in the first year: Beyond all expectations*. ProQuest Ebook Central. http://ebookcentral.proquest.com/lib/wpi/detail.action?docID=5649597

Sheller, M. (2020). *Island futures: Caribbean survival in the Anthropocene*. Duke University Press.

Stokes, F. (2023, February 20). Thousands gather at Agrifest for food, fun and flora. *The Daily News*. https://www.virginislandsdailynews.com/news/thousands-gather-at-agrifest-for-food-fun-and-flora/article_09f4ac2e-34b8-5b0e-8c68-e3cd6e070813.html

Sultana, F. (2022). The unbearable heaviness of climate coloniality. *Political Geography, 99*, 102638.https://doi.org/10.1016/j.polgeo.2022.102638

United States Virgin Islands population 2025. (2025, September 1). World Population Review. https://worldpopulationreview.com/countries/united-states-virgin-islands

U.S. Department of State. (2009). *Purchase of the United States Virgin Islands, 1917*. https://2001-2009.state.gov/r/pa/ho/time/wwi/107293.htm

U.S. Virgin Islands Department of Agriculture. (2021). *Virgin Islands Agricultural Plan*. https://doa.vi.gov/wp-content/uploads/2022/05/Virgin-Islands-Agricultural-Plan_2021-VI-Ag-Plan-Task-Force.pdf

U.S. Virgin Islands Department of Planning and Natural Resources. (2024). *The Comprehensive Land and Water Use Plan*. https://static1.squarespace.com/static/6328d6d2935e24215d41158b/t/675898bff579c04786668528/1733859531255/USVI+Comp+Plan_updated+120924+-+reduced+size.pdf

U.S. Virgin Islands Office of Disaster Recovery. (2025). 2017 hurricanes. https://www.usviodr.com/irmaandmariafunding/

U.S. Virgin Islands Department of Tourism. (2025, May 5). *St. Thomas Carnival village 2025 delivers unforgettable week of culture, music, and star power*. Visit USVI. https://www.visitusvi.com/press-release/st-thomas-carnival-village-2025-delivers-unforgettable-week-of-culture-music-and-star-power/

U.S. Virgin Islands Economic Development Authority. (n.d.). *Vision 2040*. USVIEDA. https://usvieda.org/our-vision-2040/

Virgin Islands Territorial Emergency Management Agency. (2024). *2024 Hazard Mitigation and Resilience Plan*. https://hazardmitigationplanvi.org/#flipbook-15/1/

Wang, J. (2025, March 14). Equal enfranchisement: Extending complete voting rights in the U.S. Territories. *California Law Review*. https://www.californialawreview.org/print/equal-enfranchisement#_ftn10

CHAPTER 5

The Limits to Adaptation: Reflections from VCAPS Exercises in Two Southern Coastal Communities in Puerto Rico

Sarah Molinari and Seth P. Tuler

Abstract This chapter presents lessons from two case studies where researchers engaged in the Vulnerability, Consequences, and Adaptation Planning Scenarios (VCAPS) process in two communities in southern coastal Puerto Rico as part of Whole Communities Resilience Planning Program (WCRP). We discuss how listening to participants' lived experiences of displacement and government disinvestment "unsettled" the VCAPS approach and revealed nuances around the *limits to adaptation*, the importance of place-based experiences, local flood perceptions, and everyday adaptations mediated by institutions and power relations. As a

S. Molinari (✉) · S. P. Tuler
Department of Integrative and Global Studies, The Global School, Worcester Polytechnic Institute, Worcester, MA, USA
e-mail: smolinari@wpi.edu

S. P. Tuler
e-mail: stuler@wpi.edu

S. Molinari and M. Sheller (eds.), *Building Equitable Climate Adaptation Partnerships in the US Caribbean*,
https://doi.org/10.1007/978-3-032-15727-0_5

tool of participatory research, we discuss how VCAPS became a platform for thinking about complex local hazard perceptions and participatory methodologies for climate adaptation research. We conclude by offering insights into building more equitable climate adaptation partnerships that can: (1) orient toward climate justice concerns, (2) address multiple hazards in a changing climate, (3) account for everyday adaptations and coping mechanisms that support people's desire to remain in place, and (4) give space for participants to express a broad range of community concerns and explore—rather than assume—their link to climate risks.

Keywords Climate adaptation · Everyday adaptation · Participatory methodologies · Flooding · Southern coastal Puerto Rico

1 Introduction

Caribbean islands bear disproportionate burdens from climate change and have become central focal points for climate resilience and adaptation planning (Laurent & Duvat, 2024; Rosenberg, 2020; Thomas, 2019; Thomas et al., 2019). In the US Caribbean, recent experiences of natural disasters such as Hurricanes Irma and Maria in 2017, Hurricane Fiona in 2022, and the intensification of extreme heat and flood events have both spurred and shaped planning initiatives. In December 2024, the US Virgin Islands (USVI) legislature adopted the Comprehensive Land and Water Use Plan, which sets out five to ten years of implementation goals around land and water use, natural resource protection, and sustainability (USVI Comp Plan, 2024). In Puerto Rico, the Mitigation, Adaptation and Climate Change Resilience Plan is awaiting legislative approval, and the parallel, federally funded Whole Communities Resilience Planning Program (WCRP) is underway. More recently, the Caribbean Collaborative Action Network (CCAN), a NOAA Climate Adaptation Partnership team based in Puerto Rico and the USVI, has been actively connecting university, government, and community partners to better prepare for climate risks, including flooding and extreme heat.

The record of planning and implementation in the region is mixed. In a review of 89 Caribbean Island planning documents, Thomas et al. (2019) found that they often fail to follow best practices. Challenges include weak institutional and community capacities, conflicting goals between actors and policy, lack of funding, weak integration of climate hazards, impacts, vulnerability, and risk; data gaps, top-down planning, and other more immediate concerns, such as food, housing, and employment insecurities (Dulal et al., 2009; Thomas et al., 2019). These challenges have also undermined the ability of community members to engage effectively in adaptation and resilience planning and implementation. While planners are often exhorted to create opportunities for meaningful community engagement, it has proven difficult in practice. Montoute et al. (2019) observed in a study based in Trinidad and Tobago that civil society organizations have had little influence on formal planning, while engaging mainly in activities to build awareness, training, and research. Others have noted that justice and equity concerns—spanning distributional, epistemic, historical, and structural inequalities—are not always forefronted in initiatives that tend to center technical enhancements backed by scientific expert knowledge (Araos et al., 2021; Juhola et al., 2022; Meerow et al., 2019). Furthermore, effective "climate communication"—ensuring that climate terminologies and action plans relate to impacted people's lived experience—has been challenging for ordinary Caribbean people and for partnerships engaging this work (CANARI, 2009; Moser, 2014).

In response to such challenges, emphasis is being placed on the development and application of frameworks for engagement that emphasize eliciting and organizing local knowledge and methods to integrate with and complement scientific knowledge (Iwama et al., 2021; Mercer et al., 2012). In this chapter, we explore our experience using the Vulnerability, Consequences, and Adaptation Planning Scenarios (VCAPS) process in two communities in southern Puerto Rico that participated in the WCRP program. VCAPS is a dialog-based tool for engaging community residents, planners, and researchers in revealing and understanding place-specific climate hazards and how to manage them (Clifford et al., 2023; Kettle et al., 2014; Tuler et al., 2020).

Building on the use of VCAPS in more than 20 communities in the continental US and Puerto Rico (Climate Change Council Puerto Rico, 2011; Tuler et al., 2020), CCAN researchers—led by co-authors Molinari and Tuler and CCAN researcher Laura Cabrera—were invited by the Foundation for Puerto Rico, the lead entity organizing the WCRP and a CCAN collaborator, to adapt VCAPS to address flooding hazards in two southern coastal communities of La Parguera, Lajas, and Playa de Ponce, Ponce. Our aim was to support adaptation and resilience planning by centering local knowledge and concerns and to leverage the exercise for the development of a risk and vulnerability analysis within the WCRP program. However, our experiences in fact "unsettled" the VCAPS approach in ways that led to productive methodological and practical insights. VCAPS became a platform for thinking about complex local hazard perceptions, more equitable partnerships, and participatory methodologies for climate adaptation research.

In what follows, we discuss the CCAN partnership that facilitated the VCAPS exercises and our experiences in La Parguera, Lajas, and Playa de Ponce, Ponce. The exercises revealed nuances around community experiences of the *limits to adaptation* and the importance of place-based experiences and perceptions of flooding, both of which we understand as inseparable from underlying structural inequalities and conditions of precarity impacting southern coastal communities. Reflecting on the disconnections between our VCAPS methodological approach and the lived experiences of participants, we conclude by offering insights into building partnerships that can: (1) orient toward climate justice concerns, (2) address multiple hazards in a changing climate, (3) account for everyday adaptations and coping mechanisms that support people's desire to remain in place, and (4) give space for participants to express a broad range of community concerns and explore—rather than assume—their link to climate risks.

2 Resilience Planning in Puerto Rico and VCAPS

In Puerto Rico, the physical and social devastation in the wake of Hurricanes Irma and Maria opened debates about *autogestión* and government failure, infrastructural fragility and privatization, and "just recovery"—how to rebuild in locally responsive ways that avoid reproducing existing inequalities and vulnerabilities (Ayuda Legal Puerto Rico, 2019; Molinari, 2022; Rodríguez Soto, 2020). In 2018, the US Department of Housing and Urban Development allocated $18.5 billion in Community Development Block Grants-Disaster Recovery (CDBG-DR) funds to Puerto Rico to be managed by Puerto Rico's Department of Housing. These funds cover long-term planning activities around housing, health, infrastructure, the environment, education, and economic development.

The Whole Community Resilience Planning Program (WCRP) is one of the largest CDBG-DR efforts, with $55 million designated to supporting community resilience planning initiatives that prioritize the needs of beneficiary communities (García et al., 2019). Notably, the program does not include funding for *implementation* of the resilience plans. This implementation gap tracks for the wider Caribbean, where research shows that adaptation measures tend to focus more on assessments, training, and pilot projects rather than actual implementation (Thomas, 2025).

The Puerto Rico Department of Housing contracted the nonprofit organization Foundation for Puerto Rico (FPR) to support the WCRP program with technical assistance, participatory planning, and outreach. Twenty "subrecipients" (i.e., municipalities and nonprofit organizations) were selected through a competitive proposal evaluation process to work with seventy-eight communities across the archipelago to develop locally responsive plans and strengthen the resilience of high-risk communities.[1] The WCRP planning framework is based on four phases to (1) describe current conditions, (2) complete a vulnerability and risk analysis, (3) prioritize resilience actions, and (4) implement, monitor, and, if needed, adapt the plan. Subrecipients work with communities over a nine to twelve-month period through the four phases (Puerto Rico Housing Department, 2024, p. 9).

Resilience planning is explicitly linked to climate adaptation in the WCRP framework, providing an opportunity for CCAN partners to collaborate through the VCAPS methodology and reflect on the nuances, shortfalls, and possibilities within adaptation partnerships. After several meetings with FPR program facilitators to demonstrate VCAPS, we were invited to work with one WCRP subrecipient and two local working committees. Our objective was to facilitate VCAPS to inform the Vulnerability and Risk Analysis (Phase 2) in the WCRP process, where participants identify and prioritize their resilience needs. We did not set out to assess the entire planning process, but only whether a VCAPS methodology could be helpful in one phase of the process and more broadly to better understand how local communities assess their own vulnerabilities in relation to climate risks. Thus, we did not participate in subsequent phases, or the final plan itself.

In the wake of recent disasters, the dynamics of resilience planning in the US Caribbean invite critical examination of the concept itself.

[1] See the WCRP Transparency Portal: https://recuperacion.pr.gov/wcrp/. Several entities have raised concerns over FPR securing a no-bid contract in 2019 with the Puerto Rico Department of Housing to support the WCRP program, as well as their lack of trajectory in community planning (Wiscovitch Padilla, 2019).

Caribbeanist scholars have problematized conversations around neoliberal "resilience" as casting the uneven impacts of climate hazards as a "technical problem" with depoliticizing effects that reinforce and privilege self-sufficiency and the status quo (Grove, 2014; Lloréns, 2018; Moulton & Machado, 2019; Rhiney, 2018). Others have critiqued the mobilization of resilience discourse in Puerto Rico for its individualizing impact that obscures structures of power and institutional neglect. Zorimar Rivera Montes asserts that resilience in Puerto Rico has become a "sour joke" because the discourse "discourages critical interrogation of the conditions and structures that produce adversity and instead valorizes the painful experience of surviving calamity as a personal virtue…Whoever is not individually resilient has failed" (Rivera Montes, 2025). Taking these critiques seriously invited us to reflect on the role of our participatory methodologies. The VCAPS process revealed how power and institutions map onto everyday coastal people's experiences in ways that may limit the possibilities of adaptation. To reckon with these limitations, we argue that discussions and initiatives around resilience and adaptation need to be oriented toward transformative climate justice rather than maintaining the status quo (Baptiste & Rhiney, 2016; Juhola et al., 2022; Moulton & Machado, 2019; Thomas, 2025).

VCAPS draws on the intellectual history of hazard management (Clark et al., 1998; Kates et al., 1985), climate vulnerability assessment (Dow & Carbone, 2007; Kasperson et al., 2005; Smit & Wandel, 2006), and analytic-deliberation (National Research Council, 1996) to support community adaptation by:

- Informing discussions and promoting learning by integrating climate and related natural and social sciences with local knowledge;
- Summarizing the information, knowledge, and experience that exists within a community;
- Facilitating exploration of local complexities and uncertainties;
- Stimulating discussions about how to manage consequences; and
- Aiding in future community-level decision making.

Figure 1 summarizes the phases in the VCAPS process (See Kettle et al., 2014 for more details). The first and third phases are fairly straightforward. The second phase—scenario-building—involves participants discussing, exploring, and learning about climate change-related risks, vulnerabilities, and adaptation strategies. With the support of a facilitator, participants build a "scenario" that links a selected climate hazard and place (i.e., flooding in La Parguera and Playa de Ponce) and invites participants to trace causal "pathways" through their daily lived experience and positionalities that emerge from vulnerabilities across scales. Integrating information about local interactions between biophysical and social contexts is key to understanding local phenomena in social-ecological systems so that competing priorities and values are represented and can be appropriately balanced (Picketts et al., 2012). VCAPS participants are encouraged to identify how pathways that link hazard events to consequences emerge in the local context and discuss tradeoffs among strategies of protection, accommodation, and retreat that can be implemented in different time and spatial scales. Part of the VCAPS methodological strength lies in the dialogue's potential to center experiential, place-based knowledge within climate hazard scenarios (Fig. 1). In addition, discussion can reveal and organize participants' local knowledge, allowing them to share experiences and deepen their understanding of climate risks and vulnerabilities. Discussing climate hazards from lived experience helps to make technical and abstract concepts more understandable.

FPR staff and the WCRP subrecipient organization identified participants from the working groups in La Parguera and Playa de Ponce, who themselves identified flooding as the major climate stressor to address. We adapted VCAPS to be more accessible and interactive. Following Clifford et al. (2023), we used color-coded post-it notes on a whiteboard to represent the information shared by participants rather than using a more technology-intensive approach with a computer and projector.

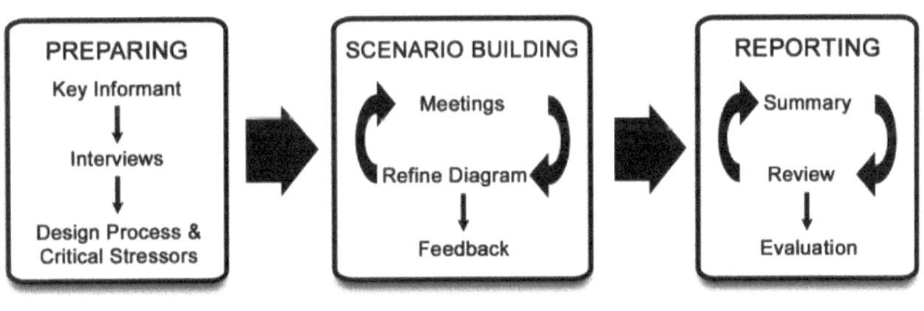

Fig. 1 Schematic of the three phases of a VCAPS process

Afterward, facilitators administered an evaluation survey to participants, created a digital visualization of the dialogue, and wrote reports for the WCRP subrecipient and working committees. In what follows, we describe the VCAPS exercises in the municipalities of Lajas and Ponce and conclude with how VCAPS became a platform for thinking about adaptation limits and participatory methodologies within equitable climate adaptation partnerships.

3 VCAPS in Two Southern Coastal Communities

Sarah Molinari and Laura Cabrera facilitated VCAPS sessions in La Parguera, Lajas, in October 2024 and in Playa de Ponce, Ponce, in November 2024. The participants from the local WCRP working committees included six full-time, elderly residents in La Parguera and three women community leaders over 50 years old in Playa de Ponce representing two neighborhoods with distinct experiences: Puerto Viejo and Villa del Carmen. The exercises led to important insights about flooding, adaptation limits, and collaborative partnerships. While we organized the sessions with certain assumptions about flooding scenarios, participants directed the conversation to urgent, albeit unexpected, issues and unmet needs that fell outside our initial framing of flood risk. Through the dialogue process, participants revealed how local flood risk perceptions are shaped by the uneven geographies of daily life in La Parguera and full-time residents' struggles around touristic overdevelopment, accessibility, and displacement. In Playa de Ponce, flood perception is shaped by government neglect, environmental injustice, and failing infrastructures. The nuances of participants' experiences in each session highlighted the centrality of "autonomous" or "everyday adaptations"—the "small, incremental changes made in our daily lives to accommodate the shifting ecologies in which we live" (Castro & Sen, 2022, p. 1). These creative everyday adaptations similarly shed light on the limits to large-scale adaptation in a context of governance failures and the importance of thinking about adaptation through a historically grounded, climate justice framework. In what follows, we describe each of the VCAPS sessions and reflect on lessons gleaned from the experience (Fig. 2).

Fig. 2 Google My Maps image of Puerto Rico marking the locations of La Parguera at the left point and Playa de Ponce at the right point

3.1 *La Parguera*

La Parguera prominently figures as a southwest paradise destination for both internal and external tourism. Historically characterized by small-scale agriculture, fishing, and a population of rural workers for the surrounding sugarcane areas, La Parguera was promoted as a tourist destination beginning in the 1950s (Valdés-Pizzini & Schärer Umpierre, 2014). Today, search results on Puerto Rico's official tourism website boast about La Parguera's unique natural attractions and water adventures, including boat rentals to explore the intricate cays and mangroves, snorkeling through the coral reefs, fishing, and tours of the bioluminescent bay (Discover Puerto Rico, 2025). Scholars have identified that tourism as an economic development strategy in climate-vulnerable islands can have adverse and extractive impacts on local communities and ecologies, exacerbating both climate and mobility injustices (Gössling & Hall, 2007; Sheller, 2021; Sheller et al., 2026). These consequences have manifested in La Parguera where paradise can bring turbulent consequences for full-time residents.

La Parguera is a dynamic socio-ecological system undergoing constant transformation. Changes in the built environment from 1930 to 2010 showed buildings and water access infrastructures increased from 130 to 1850 (López-Marrero, 2016). La Parguera has long appealed to seasonal residents and vacationers, iconically symbolized by the *casetas* ("water houses") along the sensitive coastline. Built by fishermen dating back to the 1930s, *casetas* have since become desirable for people whose livelihoods do not depend on the coastline.

Legal, environmental, and social controversies have surrounded La Parguera's *casetas* for decades. For instance, Rima Brusi conducted an ethnographic study and found that "only three out of more than 200" illegally constructed *casetas* on the coastline were primary residences, contributing to the "elitization of the coast" vis à vis "deluxe squatters" (Brusi-Gil de Lamadrid, 2008; Brusi, 2009). While *casetas* are often not recorded in the Property Registry, they tend to sell for higher prices than the average home, and owners have strategically avoided eviction over time.

VCAPS participants alluded to a recent high-profile case around illegal constructions in La Parguera. In one of the first official acts of her administration, Puerto Rico's Governor Jennifer González's nominee for Secretary of the Department of Natural and Environmental Resources signed an administrative order essentially legalizing the *casetas* in La Parguera that exist on public domain property within the Maritime-Terrestrial Zone (Méndez-González, 2025).[2] This move—denounced as illegal by environmentalists, legislators, and activists—directly benefits González's in-laws who own a *caseta* in La Parguera and have allegedly destroyed mangroves to expand the home. Furthermore, the administrative order threatens coastal communities throughout the archipelago because it sets a precedent for granting concessions to other illegal structures within the Maritime-Terrestrial Zone.

The production of leisure and touristic spaces in Lajas continues to displace full-time residents, exacerbate climate hazards, and create daily burdens around access, mobility, and quality of life. The WCRP community profile offered initial insight into infrastructural conditions exacerbating flood hazards, including ravines that clog with strong rainfall and wastewater overflows during blackouts. However, participants identified *touristic overdevelopment as the main factor that exacerbates flood risk.* One resident named Juan—a[3] lifelong fisherman—described La Parguera's "accelerated economic development" characterized by

[2] The Maritime-Terrestrial Zone is a legally-designated coastal public domain extended to where the waves reach during average storms. New legislation is proposing to change this definition to essentially reduce the public domain area and allow constructions closer to the shoreline.

[3] All participant names are pseudonyms.

construction projects without permits that harm the ecosystem. Residents shared that they sometimes conduct a kind of citizen monitoring and request to see permits of dubious constructions, often to no avail. Local ecological knowledge shapes participants' concern that the allegedly illegal constructions produce excess sediments that could damage the mangroves and crab species during flooding. In other words, illegal constructions threaten the means of local subsistence for fishermen like Juan.

Full-time residents experience a distinct relationship to La Parguera compared to absentee owners, part-time seasonal residents, and tourists. Margaret, another participant, described: "As a community, we are few because unfortunately it is a tourist area, and we have many people who have sold their properties…and those of us who remain who are from here, we are the minority, and the majority are those who are looking to profit from everything that is around us." For those who remain, the tensions between the hollowing out of full-time residents and touristic saturation create daily obstacles. She further described the hassles caused by the growing number of trailers and boats stationed along residential streets that limit access and parking availability: "Access on my street is minimal…when I get to my house, I have to look on one side where there are cars and trailers with their boats, and the other side is the same."

Flooding was not the most salient issue for participants, and their urgent concerns were often not directly linked to climate hazards. However, the VCAPS dialogue allowed participants to verbalize connections that underscore the relationship between touristic-driven spatial and economic development and increased flood risk. For instance, the trailer and boat congestion limits accessibility along streets, evacuation routes, docks, and waterways, exacerbating mobility issues for both residents and visitors in a flooding event. Participants proposed mitigating this congestion through designating more one-way streets and establishing special parking permits for residents.

Despite the obstacles, residents affirmed their desire to remain in place. "The coast is ours," as Juan put it, emphasizing that despite efforts to privatize the coast through illegal *caseta* constructions and private docks, residents defend their right to water access and public infrastructure. This was particularly salient to Juan, who makes a living selling fish in *Villa Pesquera* (the Fishermen's Village). Despite objections from the local Fisherman's Association, the municipality of Lajas is currently seeking proposals to rent Villa Pesquera's dock and kiosk facilities for recreational use, which threatens to displace small-scale fish vendors (Cotto, 2025). In this challenging context, participants spoke more of "autonomous" or "everyday adaptations" (Castro & Sen, 2022) than systematic climate mitigation efforts. For instance, fishermen in Villa Pesquera organize autonomous social infrastructures and mutual aid efforts that offer coping mechanisms in case of emergencies, or shared resources in "normal times." Juan described how local fishermen have equipped Villa Pesquera with solar energy installations for public use to charge devices, refrigerate medicines, and distribute ice—services he regularly advertises on his Facebook page in hopes that they reach people in need.

3.2 *Playa de Ponce*

Playa de Ponce is situated within a similar socio-geographic space within Puerto Rico, which Lloréns calls the "island's southern hinterland" (2021, p. 68) characterized by multiple forms of marginalization in relation to the San Juan metropolitan area and a historical concentration of contaminating and extractive industries. The racialized history of the plantation economy reverberates to the contemporary forms of environmental harm that "playeros" (residents of Playa de Ponce) live with—an expression of the "overt and covert, episodic and creeping" ways that the "coloniality of climate" is experienced (Sultana, 2022, p. 4; Sheller, 2025). Government abandonment is directly tied to flood risk in Playa de Ponce: Its low-lying geographic location at the edges of an urban watershed and intersected by three rivers exposes the area to urban runoff, river, and coastal flooding, as well as flooding caused by ineffective and poorly maintained water management infrastructure. Flood vulnerability directly shaped Playa de Ponce's population decline, from about 20,000 residents in the 1970s to now about 11,400 (Caquias Cruz, 2023).

All participants agreed that wastewater flooding in Puerto Viejo poses the most significant flood risk. For instance, during heavy rains stormwater runoff mixes with wastewater and overflows the streets with contaminated water to the extent that "all the sewers look like fountains," said Elsa, a local leader in Puerto Viejo. Contaminated water even floods the streets of Puerto Viejo on sunny days due to ineffective storm drains and pumping systems, a problem bearing significant public health risk that residents have experienced for over a decade (Román, 2022). Elsa relayed how people cannot leave their homes during these flooding events because of the foul smell and street contamination. Despite multiple government proposals for relocation as an adaptive solution, María, another community leader, affirmed that those who have stayed defend their right to remain in place. "They have been there for *centuries*," she said.

Residents of informally constructed or untitled homes in Puerto Viejo have not been afforded the same government exceptions as the *caseta* owners in La Parguera. Community leaders estimate that dozens of families left after Hurricane Maria because they were denied FEMA aid due to lack of formal property title, a problem experienced across the archipelago (Molinari, 2022). But residents began to leave the sector long before the 2017 hurricanes. In the 1990s, the government expropriated properties to construct the Port of the Americas—a deep-draft cargo and cruise ship port facility. Projects related to the Port construction turned out to be unfulfilled promises, including relocation plans for residents out of the flood zone, a dysfunctional stormwater drainage system, and a recreational court that participants claimed is useless because it has no roof or electricity. Furthermore, residents note the Port construction closed off Puerto Viejo with a fence that blocks an evacuation route in case of a tsunami (Torres Guzmán, 2018).

The VCAPS exercise revealed place-specific experiences and perceptions of flooding hazards within Playa de Ponce. In the Villa del Carmen neighborhood, just east of Puerto Viejo, flood risk is primarily related to clogged street gutters and limited accessibility. Lizette, a Villa del Carmen resident and community leader, drew a map on paper marking the neighborhood's division into three zones. The neighborhood has expanded over the years to a total of 42 streets and about 2064 homes, but has only one vehicle entry and exit—a failure of urban planning, said Lizette. She described the "chaos," panic, and vehicle congestion the limited access caused during the 2019–2020 earthquakes when residents waited in traffic jams trying to evacuate in their cars. These planning failures are accompanied by individual and municipal negligence that increases residents' vulnerability to flooding. According to Lizette, street gutters clog and cause flooding during heavy rains because of poor municipal maintenance and because "the neighbors have not cleaned the gutters. They fill up with leaves from the trees they planted themselves and they don't collect the leaves." Residents themselves have organized to find solutions to the issues of clogged gutters and evacuation routes. Similar to the Puerto Viejo experience, residents' major concerns about flooding are not related to storms or heavy rains, but rather flood risk as a socially and politically constructed experience—including poor planning decisions, dismissal of residents' concerns, and maintenance negligence.

The Port offers a kind of split-screen. For instance, while Puerto Viejo residents live amid the environmental and health harms of wastewater flooding and Villa del Carmen residents navigate street gutter maintenance and constrained access, the world's largest cruise ship docked at the Port of the Americas in May 2025. Echoing the experiences in La Parguera, the spectacle of the "Island of the Seas" arrival in Ponce was part of an effort to expand tourism to Puerto Rico's southern coast with predictable consequences.[4] Ponce's starkly contrasting realities raise questions both for tourism's impact on already-existing environmental harms and how playeros will continue to cope with new economic, social, and climate stressors.

Puerto Viejo and Villa del Carmen residents leverage extensive communication networks and personal municipal contacts to address short- and long-term problems. They have built social infrastructures that help to "fill in service gaps" (Montalto & Sheller, 2025) and support residents' coping, everyday adaptations, and self-advocacy in the face of government failure and unmet needs. Importantly, these communication and collaboration networks, neighborhood organizing, and mutual aid relations are led by women's labor and shaped by deeply rooted place-based knowledges and experiences. In other words, "*todos somos playa*," (we are all the beach) asserted María, referring to playeros' common experience and struggle despite neighborhood distinctions. The VCAPS dialogue process thus became a space for participants to articulate "embodied geographies of environmental (in)justice," which invite us to frame environmental matters as reproductive matters and attend to the gendered arrangements of environmental injustice and labors to mitigate those harms (Gay-Antaki, 2023). For instance, at the first sign of flooding, Elsa calls her Puerto Viejo neighbors to alert them to move their cars, sometimes helping elderly neighbors to do so. Then she calls her personal contacts at the public water authority and the municipality to report the issue. "When it starts to rain, I start to make calls," Elsa said.

[4] For instance, in the neighboring USVI, research shows the impact of cruise ship ports and related tourism mobilities that produce uneven geographies of access and inclusion, harm the environment, and overwhelm fragile infrastructure and local livelihoods (Sheller et al., 2026).

Similarly, Lizette messages the three WhatsApp chats for Villa del Carmen when it starts to rain to remind neighbors to check the street gutters. Women's labor in managing alerts and communications highlights the gendered dynamics of everyday adaptations in Playa de Ponce.

Community organizing, collaborations, and processes to hold institutions accountable are also key to supporting everyday adaptations and larger-scale transformations. Unlike Puerto Viejo, Villa del Carmen has a residents' association and a physical community center, assets that participants determined support self-advocacy. Board members are key to maintaining social infrastructures that mitigate flood risk and improve access and quality of life. "I personally have cleaned the gutters with a group of women—we are almost all women (in the residents' association)," Lizzette said. But residents cannot tackle unmet needs and government failures on their own. Drawing on lessons from Hurricane Maria and the 2020 earthquakes that triggered tsunami alerts, the residents' association collaborated with the *Red Sísmica de Puerto Rico* (Seismic Network) to create a pedestrian evacuation route map. Another collaboration with the Contemporary Art Museum in San Juan commissioned painted murals marking the pedestrian evacuation route (MAC, 2021). In Puerto Viejo, residents are eager to see how institutions will resolve wastewater flooding. In January 2025, after years of advocacy and organizing to hold the government accountable, the municipality of Ponce reached a Consent Decree with the Environmental Protection Agency (EPA) and the Department of Justice to address violations to the Clean Water Act for wastewater sewer overflows affecting streets, homes, and coastal waters (Environmental Protection Agency, 2025). Over the next few years, it remains to be seen how the Consent Decree actions and a completed WCRP plan might improve playeros' lives and infrastructures.

4 Reflections on Climate Adaptation and Building More Equitable Partnerships

The VCAPS exercises in two southern coastal communities in Puerto Rico generated insights around climate adaptation and partnerships. The experiences attuned researchers to the ways that community-scale adaptation can be limited by meso and macro-scale factors. Additionally, the experiences led us to re-think our own methodological tools and the role of partnerships in fostering adaptation and resilience planning. In what follows, we outline these insights and then reflect on how partnerships inclusive of government agencies, civil society groups, and local community residents with a focus on equity and justice can better address multiple hazards in a changing climate while supporting communities to pursue locally defined needs and desires.

First, governance failures and systemic, historical exclusions limit the possibilities for climate adaptation and shape how people experience and prioritize climate-related stressors. The daily concerns and everyday adaptations that participants expressed unsettled our initial conceptions of coastal "flooding" and "adaptative capacity" and thus challenged us to craft more nuanced understandings of these hazards—and the ways that local residents perceive and experience them. For instance, La Parguera residents understood flooding hazards as secondary to how they experience the daily burdens of tourism overdevelopment. Participants expressed their views about burdens related to the prevalence of illegal constructions that cause ecological harm; mobility and access constraints; the threats of displacement; and the undermining of traditional subsistence economies. In Playa de Ponce, Puerto Viejo residents described how they have coped with years of wastewater flooding due to government negligence around planning, infrastructure, and maintenance. Here, flooding from storms was not as salient a risk as the wastewater flooding caused by the infrastructure that the current EPA Consent Decree is meant to address.

In both communities, the extent of unmet needs and people's exclusion from decision-making processes challenged participants' ability to think about future-oriented adaptive solutions with the government as a trusted partner. The VCAPS dialogue process was intended to facilitate a space where residents in La Parguera and Playa de Ponce could consider how they might adapt to the challenges posed by existing governance and how to actually make practical improvements happen. In doing so, the VCAPS process helped to reveal how power, institutions, and the legacies of history map onto everyday people's experience and relationships to place in these two communities. These factors also pose structural limits to climate adaptation and resilience, shape the emergence of climate hazards, and limit the opportunities for individual technical solutions (see Thomas, 2025). This begs a question for researchers and practitioners alike to reflect on exactly *what people are expected to adapt to.* Is the emphasis on adapting to the impacts from a changing climate or the impacts of changing conditions, such as increased tourism, that may further exacerbate climate risk?

The case studies expanded our imaginary about how the changing climate impacts everyday life and decisions, and how institutions, power relations, and social relationships mediate climate risk. Furthermore, the VCAPS exercises highlighted the need for researchers to pay close attention to the ways that discussions of adaptation can become focused on individual and household actions at the expense of considering underlying structural forces that place the burden on local actors. As the VCAPS dialogue process showed, historically and socially produced climate risk does not necessarily have individual technical solutions (see Thomas, 2025).

Second, seeking pathways for resilience requires a two-sided approach to reach nuanced and inclusive understandings of the ways in which risks emerge and the potential for their management. In combination with attention to underlying structures and historical conditions, researchers and practitioners should take seriously *everyday adaptations and coping mechanisms* that build on local knowledge, histories, and place-based experiences in order to reach nuanced, relevant, and inclusive understandings of adaptation. Importantly, as Castro and Sen note, "considering *everyday action as adaptation* broadens our understanding of adaptation by allowing for the incorporation of agnostic adaptation action into adaptation research and scholarship, where hyperlocal actions in response to local ecologies aggregate into adaptive practices deeply rooted in local knowledge and society" (2022, p. 7, our emphasis). Residents of La Parguera and Playa de Ponce emphasized how their everyday adaptations are embedded within larger struggles around displacement and disinvestment.

As revealed in the two case studies, individual actions constitute important ways of managing hazards in place. The two cases also demonstrate that residents, and specifically elders, desire to remain in place and take actions to do so. At the same time, there are barriers to those who might consider moving, including lack of affordable housing and similar environmental stressors elsewhere. This holds lessons for managing climate risks and creates a temporal shift. Rather than planned, institutional adaptations with a future-orientation, a focus on everyday adaptation draws attention to people's agency and already-existing forms of coping. For example, in La Parguera, the fishermen in Villa Pesquera are confronting displacement while defending their dock and vending space, as well as the mutual aid hub that offers basic services to cope with infrastructural failures and climate threats. In Playa de Ponce, extensive communication networks, creative collaborations, and women's advocacy shape the gendered social infrastructures that residents rely on for flooding alerts, the creation of new evacuation routes, street gutter cleaning, and efforts to hold institutions accountable. Ponce's Consent Decree with the EPA to resolve wastewater flooding in Puerto Viejo is just one result of these everyday adaptations, often led by women, that support residents' abilities to remain in place. The case studies further reveal what might be lost if residents leave—the very social infrastructures and capacities that enable the individual actions supporting community resilience.

Third, our experiences with VCAPS in La Parguera and Playa de Ponce became a platform for thinking about participatory methodologies and the role of partnerships to support building adaptive capacities oriented toward more climate-just futures (see Grove, 2014). We approached the communities through the WCRP framework, which led us to frame discussions around a set of assumptions about the nature of flood risk and potential adaptive capacities in communities. It quickly became apparent in the meetings that this did not create a space for discussing the hazards most salient to participants. Their notion of resilience captures a broader range of hazards than those associated with climate change. Facilitators responded by reorienting the discussion to meet participants "where they were at," which channeled the exercise as a space for dialogue, recognition, stories, and frustrations. Being responsive to participants allowed the process to reveal and address their major concerns and unmet needs, and then make the connections to climate where possible—but not necessarily as the starting point.

Lastly, the two case studies suggest that understanding the context for everyday adaptations and coping mechanisms within broader systems can leverage opportunities for local residents to exercise agency over climate risks as they seek to remain in place and care for themselves and their communities. In discussing the urgency for genuinely inclusive forms of adaptation in the Caribbean, Rhiney and Baptiste note "that the challenge therefore is to foster and make known new ways of being that are attentive not only to future climate uncertainties and shocks, but also to the various forms of modern-day violence, injustices and inequalities that shape the day to day realities of millions of people across the Caribbean" (2019, p. 74–75). To be more effective and lead to more just outcomes, participatory methodologies engaging climate adaptation, such as VCAPS, can look toward how people understand their life conditions as the *starting point*, rather than beginning with a narrowly defined climate hazard with technical adaptive solutions. Furthermore, strategies for resilience can be further improved by considering the nuanced understandings of climate hazards embedded within institutional failures, histories of exclusion, and people's everyday adaptations and creative coping mechanisms. In research partnerships, this requires de-centering a solutions-oriented process and fostering spaces for intimate knowledge production, recognition, and storytelling to re-center people's own constructions of meaning around climate and other stressors.

The WCRP framework was not intended to play this role. Its goals are more immediately pragmatic to facilitate adaptation and resilience planning across Puerto Rico. Our experience working in this context, however, allowed us to reflect on the role that tools like VCAPS can play in resilience and adaptation planning and on the role that we play as researchers working in support of community climate adaptation.

5 Conclusion

This chapter reflects on applications of the Vulnerability, Consequences, and Adaptation Planning Scenarios (VCAPS) process as part of resilience planning in two communities in southern coastal Puerto Rico. The cases reveal the ways in which forces outside of the community structure and limit opportunities for adaptation. Inadequately designed and maintained infrastructure and incentives for tourism development are just two of the driving forces that contribute to the character of hazards in the communities. While climate stressors may exacerbate these hazards, the connection to climate is not at the forefront of local residents' concerns. The efforts to support resilience planning within a Puerto Rico-wide planning framework allowed us to reflect on the roles of participatory planning tools for climate adaptation and collaborative planning partnerships. While governance failures and systemic, historical exclusions limit the possibilities for climate adaptation, the cases also reveal the capacities local residents have for "everyday adaptations" to manage multiple hazards and remain in place.

Climate adaptation and resilience planning can be better served by learning from and leveraging such everyday actions, but without overemphasizing reliance on the depoliticizing effects of self-sufficiency. Doing so requires a reorientation away from a narrow focus on climate and toward a more expansive approach to justice and transformation. A climate justice framework is needed to facilitate transformational adaptation and partnership building. Through the identification of underlying structures and racialized power relations that produce vulnerability to climate change in the Caribbean, transformational adaptation framed through climate justice offers a move away from "business as usual approaches to adaptation, which will only continue to reinforce the already unjust status quo" (Thomas, 2025, p. 53; see also Sultana, 2022; Sheller, 2025). Centering a justice-oriented perspective on climate adaptation means fostering spaces of intimate knowledge production that value and take seriously local knowledge, connections to place, history, and stories because these factors are part of how people construct meaning and actions around climate stressors.

Competing Interests This study was partially supported through the Caribbean Collaborative Action Network (CCAN), a NOAA Climate Adaptation Partnerships team. Grant Number NA22OAR4310545.

Ethics Approval Ethics approval was granted by the Ethics Committee of Worcester Polytechnic Institute. Informed consent to participate and to publish was obtained from all individuals involved in the study.

References

Ayuda Legal Puerto Rico. (2019). *Towards a Just Recovery.*

Araos, M., Jagannathan, K., Shukla, R., et al. (2021). Equity in Human Adaptation-Related Responses: A Systematic Global Review. *One Earth, 4*(10), 1454–1467. https://doi.org/10.1016/j.oneear.2021.09.001

Baptiste, A. K., & Rhiney, K. (2016). Climate Justice and the Caribbean: An Introduction. *Geoforum, 73*(July), 17–21. https://doi.org/10.1016/j.geoforum.2016.04.008

Brusi-Gil de Lamadrid, R. (2008). Deluxe Squatters in Puerto Rico: The Case of La Parguera's Casetas. *CENTRO: Journal for the Center of Puerto Rican Studies, XX*(2), 71–91.

Brusi, R. (2009). La Costa y La Construcción Ilegal. *Revista Marejada, 4*(1). https://issuu.com/seagrantpr/docs/prsg-marejada-vol4num1_1_/s/13385065

Caquias Cruz, S. (2023, September 22). La Playa de Ponce struggles to stay afloat. ES Periódico Noticia. Retrieved July 22, 2025, from https://anthropocenealliance.org/wp-content/uploads/2025/01/22Sept_2023_-Caquias_La-Playa-de-Ponce-lucha-por-mantenerse-a-flote_ENG_Un-Nuevo-Amanecer.pdf

CANARI. (2009). *Communicating Climate Change: A Toolbox for Local Organisations in the Caribbean*. Caribbean Natural Resources Institute.

Castro, B., & Sen, R. (2022). Everyday Adaptation: Theorizing Climate Change Adaptation in Daily Life. *Global Environmental Change, 75*(July), Article 102555. https://doi.org/10.1016/j.gloenvcha.2022.102555

Clark, G., Moser, S., Ratick, S., Dow, K., Meyer, W., Emani, S., Jun, W., Kasperson, J. X., Kasperson, R. E., & Shcwartz, H. (1998). Assessing the Vulnerability of Coastal Communities to Extreme Storms: The Case of Revere, MA, USA. *Mitigation and Adaptation Strategies for Global Change, 3*, 59–82.

Clifford, K. R., Henderson, J., McAlear, Z., Dilling, L., Duncan, B., Ehert, S., Arens, R., & Rick, U. (2023). The "Nuts and Bolts" of Doing Coproduction: Exploring Implementation Decisions in Climate Adaptation Research with Stakeholders. *Bulletin of the American Meteorological Society, 104*(4), E872–E883.

Climate Change Council Puerto Rico. (2011). Working Group Meetings and Using VCAPS as a Writing Tool. Retrieved July 22, 2025, from https://new.pr-ccc.org/management/working-group-meetings-and-using-vcaps-as-a-writing-tool/

Cotto, C. (2025). Villa Pesquera de la Parguera, tierra de nadie. Claridad. Retrieved July 22, 2025, from https://claridadpuertorico.com/villa-pesquera-de-la-parguera-tierra-de-nadie/

Discover Puerto Rico (2025). *La Parguera*. https://www.discoverpuertorico.com/profile/la-parguera/12305

Dow, K., & Carbone, G. (2007). Climate Science and Decision Making. *Geography Compass, 1*(3), 302–324.

Dulal, H. B., Shah, K. U., & Ahmad, N. (2009). Social Equity Considerations in the Implementation of Caribbean Climate Change Adaptation Policies. *Sustainability, 1*(3), 363–383.

Environmental Protection Agency. (2025). *Information for Community Presentation Municipality of Ponce Consent Decree Civil Action No. 3-24-cv-01478*. EPA. Retrieved July 22, 2025, from https://www.epa.gov/system/files/documents/2024-11/ponce_eng_consentdecree_fact-sheet.pdf

García, I., Olshansky, R. B., & Carrasquillo, D. (2019). *Puerto Rico Lurches Toward Recovery*. American Planning Association. https://www.planning.org/planning/2019/aug/puertoricorecovery/

Gay-Antaki, M. (2023). Embodied Geographies of Environmental Justice: Toward the Sovereign Right to Wholly Inhabit Oneself. *Environment and Planning e: Nature and Space, 6*(4), 2379–2399. https://doi.org/10.1177/25148486231151802

Gössling, S., & Hall, M. C. (Eds.). (2007). *Tourism and Global Environmental Change: Ecological, Social, Economic, and Political Interrelationships*. Reprinted. Contemporary Geographies of Leisure, Tourism, and Mobility 4. Routledge.

Grove, K. (2014). Agency, Affect, and the Immunological Politics of Disaster Resilience. *Environment and Planning d: Society and Space, 32*(2), 240–256.

Iwama, A. Y., Araos, F., Anbleyth-Evans, J., Marchezini, V., Ruiz-Luna, A., Ther-Ríos, F., Bacigalupe, G., & Perkins, P. E. (2021). Multiple Knowledge Systems and Participatory Actions in Slow-Onset Effects of Climate Change: Insights and Perspectives in Latin America and the Caribbean. *Current Opinion in Environmental Sustainability, 50*, 31–42.

Juhola, S. K., Kayhko, J., Heikkinen, M., Schmid Neset, T., & Tuhkanen, H. (2022). Governance and Everyday Adaptations? Examining the Disconnect Between Planned and Autonomous Adaptation Through Justice Outcomes. *Ecology and Society, 27*(4), 41.

Kasperson, J. X., Kasperson, R. E., Turner, B. L. I., Hsieh, W., & Shiller, A. (2005). Vulnerability to Global Environmental Change. In: Kasperson, J. X., Kasperson, R. E. (Eds.), *The Social Contours of Risk*. Volume II. Earthscan.

Kates, R. W., Hohenemser, C., & Kasperson, J. X. (Eds.). (1985). *Perilous Progress: Managing the Hazards of Technology*. Westview Press.

Kettle, N. P., Dow, K., Tuler, S., Webler, T., Whitehead, J., & Miller, K. M. (2014). Integrating Scientific and Local Knowledge to Inform Risk-Based Management Approaches for Climate Adaptation. *Climate Risk Management, 4–5*, 17–31.

Laurent, C., & Duvat, V. K. (2024). Addressing the Climate Adaptation Tracking Gap: An Assessment Method and Its Application to the Caribbean Region. *Regional Environmental Change, 24*(4), 147.

Lloréns, H. (2018). Ruin Nation. In Puerto Rico, Hurricane Maria Laid Bare the Results of a Long-Term Crisis Created by Dispossession, Migration, and Economic Predation. *NACLA Report on the Americas, 50*(2), 154–159. https://doi.org/10.1080/10714839.2018.1479468

Lloréns, H. (2021). *Making Livable Worlds: Afro-Puerto Rican Women Building Environmental Justice*. University of Washington Press.

López-Marrero, T. (2016). Cambio del paisaje de La Parguera entre 1930 y 2010: El caso de las estructuras construidas y los manglares. 2016. *Revista Marejada, 14*(1), 16–23.

Meerow, S., Pajouhesh, P., & Miller, T. R. (2019). Social Equity in Urban Resilience Planning. *Local Environment, 24*(9), 793–808. https://doi.org/10.1080/13549839.2019.1645103

Méndez-González, L. J. (2025, January 15). Environmental Lawyers: Order Tied to Governor's In-Laws Is Illegal, Centro de Periodismo Investigativo. Retrieved July 22, 2025, from https://periodismoinvestigativo.com/2025/01/environmental-lawyers-order-tied-governors-in-laws-illegal/

Mercer, J., Kelman, I., Alfthan, B., & Kurvits, T. (2012). Ecosystem-Based Adaptation to Climate Change in Caribbean Small Island Developing States: Integrating Local and External Knowledge. *Sustainability, 4*(8), 1908–1932.

Molinari, S. (2022). Disaster Fraud Prevention by Exclusion: Property, Home-ownership, and Individual Housing Repair Aid in Puerto Rico. *CENTRO: Journal of the Center for Puerto Rican Studies, XXXIV*(2), 327–351.

Montalto, F., & Sheller, M. (2025, June 4). Strengthening Social Infrastructure for Climate Resilience. *The Nature of Cities*. https://www.thenatureofcities.com/TNOC/2025/06/04/strengthening-social-infrastructure-for-climate-resilience/

Montoute, A., Mohammed, D., & Francis, J. (2019). Prospects and Challenges for Civil Society and Climate Change Engagement in the Caribbean: The Case of Trinidad and Tobago. *Caribbean Studies, 47*(2), 81–109.

Moser, S. C. (2014). Communicating Adaptation to Climate Change: The Art and Science of Public Engagement When Climate Change Comes Home. *Wires Climate Change, 5*(3), 337–358. https://doi.org/10.1002/wcc.276

Moulton, A. A., & Machado, M. R. (2019). Bouncing Forward After Irma and Maria: Acknowledging Colonialism, Problematizing Resilience and Thinking Climate Justice. *Journal of Extreme Events, 6*(01), 1940003.

Museo del Arte Contemporáneo de Puerto Rico. (2021). Murales villa del Carmen. San Juan: Museo del Arte Contemporáneo de Puerto Rico. Retrieved July 22, 2025, from https://www.museomac.org/arte/comisiones/Murales-Villa-del-Carmen

National Research Council. (1996). *Understanding Risk: Informing Decisions in a Democratic Society*. National Academy Press.

National Research Council. (2008). *Public Participation in Environmental Assessment and Decision Making*. National Academy Press.

Picketts, I. M., Curry, J., Rapaport, E. (2012). Community Adaptation to Climate Change: Environmental Planners' Knowledge and Experiences in British Columbia, Canada. *Journal of Environmental Planning Policy Manage, 14*(2), 119–137.

Puerto Rico Department of Housing. (2024). *Whole Community Resilience Planning Program (WCRP): Planning Framework*.

Rhiney, K. (2018). Recipe for Resilience? Tracing the Biopolitics of Sint Maarten's Recovery Efforts After Irma. *Journal of Extreme Events, 5*(04), 1840002.

Rivera Montes, Z. (2025, April 30). *It's About Time We Fail: A Puerto Rican Poetics of Exhaustion.* Public Books. Retrieved July 22, 2025, from https://www.publicbooks.org/its-about-time-we-fail-a-puerto-rican-poetics-of-exhaustion/

Rodríguez Soto, I. (2020). Mutual Aid and Survival as Resistance in Puerto Rico. *NACLA Report on the Americas, 52*(3), 303–308. https://doi.org/10.1080/10714839.2020.1809099

Román, O. (2022, January 20). Una comunidad ponceña denuncia problema de aguas usadas. *El Vocero.* https://www.elvocero.com/comunidad/una-comunidad-ponce-a-denuncia-problema-de-aguas-usadas/article_a139e9a2-7986-11ec-9d75-5b0e239acc2a.html

Rosenberg, J. (2020). Adaptation, Official Development Assistance, and Institution Building: The Case of the Caribbean Community Climate Change Centre. *Sustainability, 12*(10), 4269.

Sheller, M. (2025). Caribbean Islands and the Coloniality of Climate Change: Navigating 'the Anthropocene' Through the Historical Legacies of the Plantation. In P. Noxolo, K. Rhiney, & R. Cummings (Eds.), *Routledge Handbook of Caribbean Studies* (pp. 35–46). Routledge International Handbooks. Routledge. https://doi.org/10.4324/9781003136743.

Sheller, M. (2021). Reconstructing Tourism in the Caribbean: Connecting Pandemic Recovery, Climate Resilience and Sustainable Tourism through Mobility Justice. *Journal of Sustainable Tourism, 29*(9), 1436–1449. https://doi.org/10.1080/09669582.2020.1791141

Sheller, M., Trotman, L., Guannel, G., & Waddell, K. (2026). Tourism Mobilities, Climate Mobilities, and Mobility Injustice in the US Virgin Islands. In F. Higgins-Desbiolles, R. Rastegar, & R. K. Shrestha (Eds.), *Climate Justice in Tourism: Centring Justice and Equity in Our Responses to Climate Change.* Channel View Publications.

Smit, B., & Wandel, J. (2006). Adaptation, Adaptive Capacity and Vulnerability. *Global Environmental Change, 16*, 282–292.

Sultana, F. (2022). The Unbearable Heaviness of Climate Coloniality. *Political Geography, 99*(November), Article 102638. https://doi.org/10.1016/j.polgeo.2022.102638

Thomas, A. (2025). Transformational Adaptation to Climate Change in the Caribbean. In P. Noxolo, K. Rhiney, & R. Cummings (Eds.), *Routledge Handbook of Caribbean Studies* (pp. 47–60). Routledge.

Thomas, A. (2019). Adaptation Planning in Caribbean Small Island Developing States: A Literature Review. In C. Klöck & M. Fink (Eds.). *Dealing with Climate Change on Small Islands: Towards Effective and Sustainable Adaptation?* (pp. 123–140). Göttingen University Press.

Thomas, A., Shooya, O., Rokitzki, M., Bertrand, M., & Lissner, T. (2019). Climate Change Adaptation Planning in Practice: Insights from the Caribbean. *Regional Environmental Change, 19*(7), 2013–2025.

Torres Guzmán, S. (2018, July 11). Puerto Viejo exige acción para preservar la comunidad. *Primera Hora*. https://www.primerahora.com/noticias/puerto-rico/notas/puerto-viejo-exige-accion-para-preservar-la-comunidad/

Tuler, S., Dow, K., & Webler, T. (2020). Assessment of Adaptation, Policy, and Capacity Building Outcomes from 14 Processes. *Environmental Science and Policy, 114*, 275–282.

USVI Comp Plan. (2024). *Planning Our Future: The Comprehensive Land and Water Use Plan*. https://www.planusvi.com/

Valdés Pizzini, M., & Schärer Umpierre, M. T. (2014). *People, Habitats, Species, and Governance: An Assessment of the Social-Ecological System of La Parguera, Puerto Rico*. With Caribbean Coral Reef Institute, University of Puerto Rico (Mayagüez Campus), United States, and Centro Interdisciplinario de Estudios del Litoral. University of Puerto Rico, Interdisciplinary Center for Coastal Studies.

Wiscovitch Padilla, J. (2019, February 27). *Without a Proposal, Foundation for Puerto Rico Scores Largest CDBG-Funded Contract*. Centro de Periodismo Investigativo. https://periodismoinvestigativo.com/2019/02/foundation-for-puerto-rico-scores-largest-contract/

CHAPTER 6

Filming Community *Autogestión*: Insights from a Participatory Video Project

Bárbara López-González

Abstract This chapter reflects on the use of Participatory Video (PV) in bottom-up transformational efforts toward sustainability and the pursuit of alternative futures. Focusing on the experience of *Casa Pueblo*, a community *autogestión* organization in Adjuntas, Puerto Rico, this chapter advocates for community-engaged, context-sensitive, co-created

The Participatory Video Project was developed collaboratively with members (staff and volunteers) from *Casa Pueblo*: Laura Irizarry, Jenniffer Pérez, Claudia Irizarry, Verónica Aponte & Natalia Irizarry. While they preferred to be acknowledged for their contributions rather than listed as co-authors of this chapter, this does not diminish their contribution and involvement in co-producing the film. The insights discussed in this chapter draw on doctoral research and analysis conducted by López-González at the University of Guelph, ON, Canada.

B. López-González (✉)
Worcester Polytechnic Institute, Worcester, MA, USA
e-mail: blopezgonzalez@wpi.edu

S. Molinari and M. Sheller (eds.), *Building Equitable Climate Adaptation Partnerships in the US Caribbean*,
https://doi.org/10.1007/978-3-032-15727-0_6

methodologies that embrace multiple ways of knowing and acting. To achieve this, it draws on the Participatory Video *Voces desde la Autogestión*, an academic project under the initiative *Casa Pueblo Universitario*. This project is grounded in the lived experiences, everyday actions, and insights of the organization's members (staff and volunteers) and shaped by local contexts and participants' prior experiences. The chapter showcases how community *autogestión* can support both the deconstruction of colonial beliefs and transformations. Furthermore, it demonstrates how filmmaking can serve simultaneously as a method of research and a collective practice of critical reflection and engagement, serving as a decolonial and future-oriented practice. As a result, the chapter presents filmmaking as a tool to co-produce knowledge, engaging both academic and community expertise, positioning the researcher as facilitator rather than only expert. To watch the full video, see: https://atrium.lib.uoguelph.ca/items/e56cf12a-adeb-4381-b9a7-a025a52af51b

Keywords *Autogestión* · Transformations · Participatory Video · Puerto Rico · *Casa Pueblo*

1 Introduction

"Transformations toward sustainability require alternative socio-ecological interactions that enable a different set of conditions—in other words, alternative futures. Despite the urgent necessity of changing the current conditions, top-down, large-scale, one-size-fits-all approaches to sustainability transitions have largely failed to generate meaningful changes—often reinforcing business as usual—and have left communities with limited options to confront and overcome their ongoing and interrelated social, economic, and environmental challenges (O'Brien et al., 2023; Temper et al., 2018). In this scenario, where community members are discouraged and in need of creative and innovative options, local and context-specific initiatives are playing an important role in offering alternative pathways for communities trapped in historical cycles of injustice and vulnerability. Such pathways not only attempt to resist or offer hope but also emancipate communities from experiences of oppression and exploitation (Rodríguez et al., 2024; Stock, 2023; Temper et al., 2018).

Citizen-led, community actions play a critical role in designing creative and innovative initiatives for transformations to sustainability as well as implementing sustainable efforts in areas such as agroecology, solar energy, and water sharing networks (Bennett et al., 2016; Bentz et al., 2022; Temper et al., 2018). As complex processes that entail changes at multiple levels, transformations toward sustainability involve qualitative shifts in ways of thinking and acting that aim to address the root causes of current experiences of unsustainability, rather than merely adjusting to their impacts (Bentz et al., 2022; Pelenc et al., 2019; O'Brien & Sygna, 2013). Bottom-up efforts, such as community *autogestión* (defined below), have been able to spark these transformations by moving "across, over, and beyond the current state of affairs" (Vogel & O'Brien, 2022: 653), developing holistic and integral initiatives with the capacity to address local needs while "recreating their worlds from the ground up" (Rodríguez et al., 2024: 1). Despite their capacity to catalyze transformations, these efforts have often been dismissed or minimized in institutional and academic discussions about transformations due to concerns regarding their limited capacities and the difficulties of scaling up (O'Brien et al., 2023). These critiques present a challenge not only for citizen-led efforts seeking to make their proposals visible, but also for academics and researchers, who must consider how to embrace rather than continuing to exclude diverse voices and knowledge that can contribute to broader discussions on transformations (Bai et al., 2016; Bentz et al., 2022; Cui & Brombal, 2023; Vogel & O'Brien, 2022).

In response to these challenges, this chapter seeks to shift the discussion away from broad and abstract notions of transformation toward local perspectives and everyday actions that deepen understanding and enable transformative change, focusing on the experience of community *autogestión* in Puerto Rico (Bentz et al., 2022; Bluwstein, 2021). In doing so, the chapter aims to explore what Vogel and O' Brien (2022: 655) describe as the "heart" of transformation, drawing on the Participatory Video *Voces desde la Autogestión* developed with the members (staff and volunteers) of the organization *Casa Pueblo*. This project is grounded in the lived experiences and insights of the organization's members, offering a perspective that differs from those typically explored within the organization. In doing so, it reveals *autogestión* not only as a response to ongoing local challenges but also as an everyday practice of deconstruction of colonial beliefs and transformations. Furthermore, it presents filmmaking as a decolonial and future-oriented practice, thereby illustrating how academic and community expertise can collaborate, positioning the researcher as facilitator rather than the only expert. From the experience of this PV, the chapter advocates for community-engaged, context-sensitive, co-created methodologies that embrace multiple ways of knowing and acting.

2 Citizen-Led Efforts: Community *Autogestión*

Autogestión is commonly translated as "self-management." The term, derived from the Greek autós (self) and gestio (managing), originates in Europe, where it functions as both a countermeasure to neoliberal capitalist modes of production and a form of resistance (Hudson, 2010; Rodríguez Tamayo, 2019). The genealogy of the term is tied to nineteenth-century socialist, Marxist, and anarchist traditions, both of which emphasized the liberation of labor and the democratization of social life (Rodríguez Tamayo, 2019; Vieta, 2014). From this perspective, supporters envisioned decentralized systems of governance where workers would directly control the means of production and actively participate in shaping social structures. Although the concept is historical, the idea of *autogestión* has been appropriated over time by various ideological currents, including depoliticized interpretations associated with neoliberal reforms and state retrenchment. As a result, some contemporary uses reduce the term to self-management, overlooking its original political intent (Méndez & Vallota, 2006; Ortiz Gómez, 2013; Rodríguez Tamayo, 2019).

In Latin America and the Caribbean, regions shaped by colonialism, racial capitalism, and struggles for territorial sovereignty, *autogestión* expands beyond the management of the source of production by workers or management of one's own affairs and leans toward social transformation and an integral management of society. It is a practice rooted in postcolonial and decolonial frameworks and cannot be detached from historical socio-political processes of resistance by Indigenous people, Afro-descendants, and peasant movements who have long organized outside of state structures to sustain life, protect the territory, and transform social, cultural, and ecological relations (Escobar, 2018; Moulton, 2024; Post, 2023; Rodríguez Tamayo, 2019). In these contexts, *autogestión* emerges as a way to refuse the future offered by colonial dynamics and secure critical resources necessary for sustaining life.

In the case of Puerto Rico, scholars have connected the idea of community *autogestión* with efforts to resist capitalist exploitation

(Lloréns, 2021; Massol-González, 2022) and preserve the commons (García-López et al., 2017). Most recently, they have also explored the role of *autogestión* in the wake of hurricanes Irma (2017–Category 5) and Maria (2017-Category 4) as community-led responses to disasters (Garriga-López, 2019; Rivas, 2018; Rodríguez-Soto, 2020; Serrano-García, 2020), a "coalitional counter-praxis" (Soto-Vega, 2019), and a "strategy of survival" (Roque et al., 2021). It has been a key practice in responding to emergencies and in providing and distributing assistance, while simultaneously fostering mutual aid and solidarity in the face of a system that "reproduces death" (Avilés-Vázquez et al., 2023; Torres-Abreu et al., 2023; Massol-González, 2024; Villarrubia-Mendoza & Vélez-Vélez, 2020).

These days, *autogestión* is commonly embraced by or aspired to by many community organizations, with diverse ideologies, causes, and varying capacities and resources. These organizations, rooted in ideas of social, economic, and environmental justice, against colonial governance, and in relation to causes like gender and racial equality, are distributed throughout the Puerto Rican archipelago (Tormos, 2018). In the face of historical and ongoing exploitation and extractivism as well as constant emergencies, they embrace *autogestión* as a method to face disasters, not only as isolated moments of disruption, but as an ongoing, everyday experience where bodies and territories remain continuously vulnerable. In doing so, they transcend the reactive response to the emergencies, moving toward proactive and intentional efforts to create new possibilities—possibilities that attempt to open pathways toward alternative futures beyond the ones already prescribed (Cork et al., 2023).

3 An Illustrative Case of Community *Autogestión*: *Casa Pueblo*

Casa Pueblo[1] serves as an illustrative case of community *autogestión* in Puerto Rico. Originating in 1980 from community resistance to open-pit mining, *Casa Pueblo* has evolved from a movement of resistance, without discarding it, to a formal community organization with projects related to

[1] *Casa Pueblo* translates to "the People's House." Alexis Massol-González (co-founder) won the Goldman Prize (2002) for Islands and Island Nations in recognition of their initiatives.

nature conservation, education, culture, and solar energy. With headquarters in the center of the municipality of Adjuntas (central mountain range of Puerto Rico), they have their own facilities from which they make decisions and execute projects. These headquarters have expanded beyond the house (central space) to include a Radio Station,[2] a rescued school,[3] and other spaces such as *Bosque Escuela Ariel Massol Deyá*,[4] and *La Plaza de la Independencia Energética*.[5]

Casa Pueblo's efforts have become particularly notable after the hurricanes Irma (2017–Category 5), Maria (2017-Category 4), and Fiona (2022-Category 1) as the *autogestión* organization turned into an energy oasis and a reference point for both community-led and government-led initiatives. The organization's solar energy initiative has provided power for homes, critical infrastructure, local business, and public spaces, developing what they conceptualize as an "energy insurrection" (López-González, 2025a; Massol-Deyá, 2022). This insurrection, disruptive as it is, led from the bottom-up rather than through top-down public policy, attempts to challenge the narratives and actions of a system that promotes privatization and dependency on fossil fuels while offering "proof of concept" of the possibilities that enable alternative futures (Dunlap & Tornel, 2023; Massol-Deyá, 2022; Stock, 2023).

4 Participatory Video Project—"*Voces desde la Autogestión: Casa Pueblo*"

Participatory video (PV) is a collaborative approach to empower participants to express their experiences, perspectives, and concerns on film while actively contributing to both the creative and filming process (Kindon et al., 2007; Mitchell et al., 2018). This approach originated in Fogo Island, Canada, in the 1960s through a community partnership with the National Film Board that engaged Islanders in the creation

[2] Puerto Rico's first community and ecological radio station.

[3] Repurposed as the Community Institute for Biodiversity and Culture.

[4] Solar Forest, an educational project to humanize learning processes using the forest as pedagogical space.

[5] The Energy Independence Square is an open and participatory space that combines science, culture, and community, aiming to serve as an alternative for community members in times of emergency or need, and as a laboratory for energy research throughout the year.

of short films (Crocker, 2003). The collaborative nature of PV distinguishes the practice because participants are situated as part of the creative process rather than only "subjects" of the film (Sandoval Rivera, 2017). PV engages community members as co-filmmakers to create videos in group settings regarding themes of interest. This approach allows them to explore, question, and honor their own knowledge and experiences and share them in their "own terms" (Baldacchino, 2008).

The PV project "*Voces desde la Autogestión: Casa Pueblo*" began in May 2023 as part of an academic residency under the initiative *Casa Pueblo Universitario*,[6] a program that aims to support education and knowledge generation through partnerships with universities, researchers, and students. The project focused on the experiences of members (staff and volunteers) of *Casa Pueblo* and was designed in five phases to align with the organization's specific context and to accommodate participants' availability (see Fig. 1). This phased structure ensured flexibility, allowing the research process to adapt to the organization's ongoing initiatives and the time constraints of those involved.

Phase 1: Invitation

As part of an academic residency in *Casa Pueblo,* I met with *Casa Pueblo*'s Associate Director to discuss interest in facilitating a PV project, identify potential participants, and assess available facilities. Following

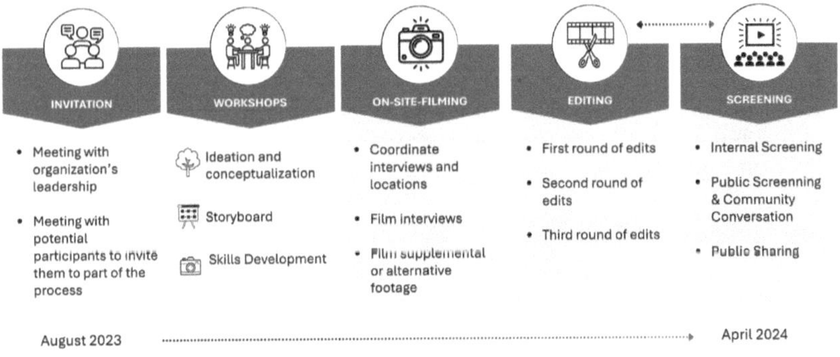

Fig. 1 Process of Voces desde la *Autogestión: Casa Pueblo*

[6] Educational initiative.

the conversation, I compiled a list of eight potential participants and extended individual invitations for an initial conversation. Four participants attended the first meeting in *Casa Pueblo*, where I introduced the idea of a collaborative video project and moderated a discussion on potential themes; participants suggested *Bosque Escuela Ariel Massol-Deyá*, *Jardín de las Mariposas,*[7] and lessons learned from the organization's co-founder, Faustina "Tinti" Deyá Díaz. After the meeting, I followed up with all potential participants to confirm their interest and time availability. Those who responded positively received an invitation to the meeting, examples of participatory videos in both English and Spanish, and the informed consent documents.

Phase 2: Workshops

The workshops began in September 2023 with a group of five women participants who were members of *Casa Pueblo*. At that time, all five women were staff or volunteers of the organization, regularly assisting with administrative tasks, group orientations, forest visits, and art classes. I led the workshops, which focused on ideation and conceptualization and incorporated art-based approaches and skill-building exercises. A total of three workshops were held centered on *Casa Pueblo*'s *autogestión*, exploring current efforts and strategies, and its role in Puerto Rico's present and future. The workshop exercises drew inspiration from the handbook *Insights into Participatory Video* (Lunch & Lunch, 2006) and were adapted to accommodate the number of participants, the available time, and place-based culture, language, and manners (see Fig. 2). The exercises included a problem tree session, a telegram-style message exercise, a storyboard creation, and practice exercises with different shot types. Each workshop session was recorded and documented for further reflection and analysis.

Phase 3: On-site filming

After the workshops, the participants and I conducted seven on-site filming sessions in four locations: *Casa Pueblo's* headquarters, *Bosque*

[7] The Butterfly Garden, inaugurated in 2001 and relaunched in 2022, serves an educational and conservation space managed by community and organization members.

Fig. 2 Problem tree exercise. Photo: Author

Escuela Ariel Massol-Deyá,[8] *Cerro Mágico,*[9] and the Public Square of Adjuntas. During the workshops, we agreed on a preliminary schedule to match scene, interviewee, interviewer, and filmmaker. There was not a script, but a set of prompts and open-ended questions based on our workshop conversations to help moderate and illustrate the diversity of people and efforts involved. We identified a total of 11 interviewees, all community members involved with *Casa Pueblo*. The interviewees were identified and the interviews recorded with the purpose of showing the spectrum of initiatives within the organization. We all participated in the process in different roles, such as filmmaker, interviewee, interviewer, and guide, providing direction on where to film and how to get there (see Fig. 3). *Casa Comunitarias de Medios*[10] also participated in the process as a collaborator, facilitating complex footage such as drone images. We held three meetings during the filming process to review footage and provide feedback. The comments from those meetings guided the filming and the analysis process.

[8] School Forest—located approximately 10-minute drive from the headquarters.

[9] Magic Hill—a camping and observation site located approximately 15-minute drive from the headquarters.

[10] Community-based media effort part of *Iniciativa de Eco-Desarrollo Bahía de Jobos* (*IDEBAJO*), another community *autogestión* organization at the community of Aguirre in Salinas, Puerto Rico.

Fig. 3 On-site filming session. Photo: Author

Phase 4: Editing

We developed the editing process collaboratively through a process of viewing-feedback-action. I managed the raw material, making a preliminary selection based on the notes from the workshops and conversations with participants (see Fig. 4). I showed those materials to the participants to seek their feedback, reflections, and guidance. After their feedback, I delivered the material to *Casa Comunitaria de Medios*, who provided technical support in the editing process, performing actions such as cleaning audio, balancing colors, managing transitions, and incorporating English subtitles. Once there was a preliminary version of the video, I reviewed it with the participants and discussed edits. Several rounds of this process took place over approximately three months, until a full draft of the film "*Voces desde la Autogestión: Casa Pueblo*" was ready.

Phase 5: Screening

Three screenings were held before the public screening at the Cine Solar.[11] The first one was in December 2023, when four of the participants reviewed the first draft of the video and provided feedback. The

[11] "Solar Cinema" at the *Casa Pueblo* headquarters in Adjuntas.

Fig. 4 Selection of scenes based on the storyboard process at the workshops. Photos: Author

second screening, in January 2024, was through Zoom with all five participants. The feedback related to the order of scenes, credits, and the name of the video. The third screening took place on Zoom with the organization's Associate Director in January 2024 to share the results and coordinate a public screening. The public screening was at the *Cine Solar* on April 20, 2024, as part of the activities for the organization's 44th anniversary (see Fig. 5).

After the screening, a conversation between members and visitors took place in which the public provided feedback, asked questions, and

Fig. 5 Public screening at *Cine Solar*. Photo: Author

reflected on Puerto Rico's current challenges and opportunities. The video also generated reflections regarding the organization's past experiences, current dynamics, and future projects. To watch the full video, see (Fig. 6):

Fig. 6 Film still from "*Voces desde la Autogestión: Casa Pueblo.*" *To watch the full video, see:* https://atrium.lib.uoguelph.ca/items/e56cf12a-adeb-4381-b9a7-a025a52af51b

5 Ethical Considerations

PV is a valuable approach for engaging with community members in the co-production of knowledge. However, it has also raised ethical concerns regarding informed consent, ownership, and power dynamics (Cardinal, 2019; Fisher et al., 2021; Marzi, 2023). To address these concerns, measures were implemented throughout the film's creation process. First, informed consent was developed in accordance with the research ethics guidelines of the University of Guelph. It was translated to Spanish and explained to participants to facilitate accessibility and understanding. Second, the final version of the film was published with credits clearly established, acknowledging individual roles and contributions. Furthermore, the film was licensed under a Creative Commons license. Each workshop participant, along with *Casa Pueblo*, received a link that can be accessed and used at any time. Lastly, although power dynamics are difficult to overcome, throughout the filming and editing process, I consistently sought to position myself as a facilitator and a listener. In doing so, I was committed to respecting diverse perspectives and integrating participants' insights into the process whenever possible. This chapter acknowledges the contributions of all participants, recognizing their intellectual effort in creating the film.

6 Analyzing Participatory Video for Insights into Transformative Change

The PV process generated drawings, pictures, audio, and video materials. In addition to the creation of the film, I transcribed a total of nine hours of material from the workshops, meetings, and on-site filming sessions and took notes documenting the filming sessions. I analyzed and coded these transcriptions and notes to shed light on how *Casa Pueblo* drives transformation from the perspective of workshop participants and PV interviewees. Several conversations during the workshop processes highlighted people's views regarding the organization and Puerto Rico. As explained by a participant:

> People have a perspective that those who, in one way or another, critically confront the status quo, whatever that may be. Oh, well, you're a communist, a socialist, an independentist, a '*machetero*.'[12]

In contrast to these catchall labels, interviewees see their efforts within *Casa Pueblo* as dynamic rather than static, having evolved over time and based on community needs rather than political ideologies. They view *autogestión* as an effort that brings together people of diverse ideologies, backgrounds, and political preferences, as stated by several project participants:

> It is an open space for all those who share the same purpose of seeking an alternative Puerto Rico to the one we have today, with a perspective of empathy, solidarity, and self-sufficiency.[13]
>
> [*Casa Pueblo*] It is much more than a museum, much more than a movement—which it is not—of a political party. It is a look at the needs of a community that has been forgotten for many years. It is a look at what it means to be Puerto Rican, at our roots, at our environment, which is our home.[14]
>
> Although the scientist is a scientist, and there is the engineer and the musician, I believe there is an element that serves as the starting point, which is that common ground, that desire to break away from the model of colonialism, dependency, and to build freedoms such as energy independence. When those points unite us, the artist knows where we collectively want to push—the engineer, the scientist, the teacher, the student. I think that helps a little in establishing a common agenda.[15]

Through the convergence of diverse people, participants facilitate counter-narratives to historical views regarding community *autogestión* and Puerto Rico's present and future. Participants described *autogestión* as an everyday process of "deconstruction" of multiple beliefs—for instance, the belief that those who protest or question the government's

[12] Popular name for a clandestine and insurgent militant organization in Puerto Rico that fought for independence over several decades.

[13] On-site filming session, Participant 1.

[14] On-site filming session, Participant 1.

[15] On-site film session, Participant 10.

actions are merely criminals. Similarly, they identified the need to challenge the commonly held view that Puerto Rico is a lost cause, a common sentiment, particularly after Hurricane Maria.

> It is a process of deconstruction at a social, political, and religious level, allowing us to carry out different actions based on necessity, which I believe is how things are done here.[16]
> During Hurricane Maria it [*Casa Pueblo*] served as an oasis, and we showed that, yes, it is very important, that it can be done, and that we do not have to continue dragging our feet while people continue to suffer.[17]

According to the participants, the organization's initiatives challenge historical colonial narratives of lack of capacity and resources, dependency, and disconnection from nature through the development of counter-narratives and actions. To do so, they embrace education as key to fostering individual and collective transformations. Participants highlighted how, in their everyday actions, they aim to provide community members and local and international visitors with experiences that educate, foster awareness, and cultivate commitment, something in which *Casa Pueblo*'s co-founder, Faustina "Tinti" Deyá Díaz, played a key role.

> Knowledge is joined with community wisdom, forming a winning equation: science + culture + community. This equation paves the way for a radical change in Puerto Rican society.[18]
> Using different approaches in this regard, education was key in conveying a message that could bring about social, political, and community changes, ensuring that every visitor became part of this.[19]
> *Casa Pueblo* has begun to foster critical knowledge in people. Rather than resigning ourselves to the idea that opportunities do not exist, it encourages us to see where those opportunities arise. This can be seen more as an education in social awareness, critical judgment, and the ability to identify opportunities that can help us.[20]

[16] Workshop, Participant 1.

[17] On-site film session, Participant 13.

[18] On-site film session, Participant 6.

[19] On-site film session, Participant 1.

[20] On-site film session, Participant 2.

Education materializes through diverse ways of acting that aim to connect, unite, and constantly engage more people in the efforts. From *Radio Casa Pueblo* to *Bosque Escuela Ariel Massol Deyá*, organizers are constantly welcoming people, making them feel part of the efforts and delivering, explicitly or implicitly, messages supported by science and intentional actions. As highlighted by the participants:

> [When it comes to artwork] It is not just a mural, but behind the mural there is a message. Just like the art workshops that I offer here at *Casa Pueblo* —behind each workshop, there is always a message about something that is happening at that moment in society, something that is there, and where change is needed.[21] (For example, see Fig. 7)

> [When it comes to distribution of aid] You saw how people from the neighborhood of Guilarte would come here to receive a solar lamp. Yes, but when you received the solar lamp, you were also educated. What is solar energy? If you have solar energy, how can you use it efficiently? And after going through that process, you took your little lamp home.[22]

Fig. 7 Artwork by the artist Danny Torres located near the town square of Adjuntas, in the parking lot of a local business. Photo: Author

[21] On-site film session, Participant 4. For more examples, see López-González (2025b). Also, see: https://casapueblo.org/murales/

[22] Workshop, Participant 1. For more example, see https://www.youtube.com/watch?v=-QH3-S4zVTM

> [Daily visits] educate students about the importance of nature and how it has an impact not only here in the centre of the island but on the entire island. It is for them to learn that water is important, that we must take care of it, that trees are there for a reason—they give us oxygen, they provide shade—and to help them make the connection between the forest and the environmental or climatic changes that are happening.[23]

Therefore, by combining education with action, such as art workshops that share critical messages and forest visits that promote a relationship with nature and with each other, *Casa Pueblo* cultivates a message of change in community members and visitors. In doing so, they constantly challenge dominant colonial views and narratives regarding community *autogestión* and Puerto Rico's lack of capacity and long-standing patterns of power resulting from colonialism and coloniality. As a result, they point to possibilities for the future, one in which people have respect and a strong relationship with one another, with nature, and with what they call *Madre Isla*.[24] This is a future in which local residents and visitors work together toward a common and collective good and the construction of freedoms.

7 Using Participatory Video

The production of *Voces desde la Autogestión: Casa Pueblo* was a rich process of a researcher working alongside community members. The initial intention was to fully engage workshop participants in all stages—interviewing, filming, editing, screening—letting them lead the entire process (Kindon et al., 2007; Mitchell et al., 2018; Whiting et al., 2018). However, several variables impacted the process. First, there was no existing group of filmmakers in place. Instead, the team had to be formed entirely from volunteers, which presented challenges in recruiting, maintaining consistency, and achieving continuity. Second, since *Casa Pueblo* hosts activities every day, the participants' workload makes it difficult for

[23] On-site film session, Participant 3.

[24] Mother Island. Concept inspired by the work of Puerto Rican thinker Eugenio María de Hostos and refers to Puerto Rico. The term also serves as the name for the organization's coffee brand. For more information, see https://casapueblo.org/product/cafe-madre-isla/

them to find time for training, multiple workshop sessions, and hands-on editing sessions, which required adjustments to the original project design. Finally, despite the organization's interest in the project, many other activities were taking place simultaneously. As a result, the project schedule had to be constantly adapted to the organization's ongoing activities and internal dynamics.

As I was completing an academic residence with *Casa Pueblo*, the daily face-to-face interactions with participants served as a good foundation to manage these variables and for the participants to get to know me, understand the research project, and be willing to participate (Leeuw et al., 2012; Morales et al., 2021). Likewise, it established a strong basis to position me as a facilitator. Considering that PV participants were volunteers and had other commitments, the idea of expecting them to be fully engaged and responsible for everything was unrealistic. In light of this, my position as facilitator helped to assume the responsibilities that PV participants were not able to assume, thus serving as a bridge between participants and collaborators (e.g., *Casa Comunitaria de Medios*). This was a "reciprocal approach" in which the PV process went in two directions (Marzi, 2023). Therefore, both PV participants and I were collaborating, learning, exchanging ideas, and assuming responsibilities, fostering a horizontal dynamic in which the roles are rotated.

This project serves as an example that there is not a "right" way to facilitate a PV. It depends on the context (e.g., time available; activities), available resources (e.g., human; economic; technical), and intended purpose (e.g., research; advocacy). However, certain principles and values should guide the process. From this experience, collaboration, reciprocity, and long-term relationship building are essential foundations. Likewise, trust, accountability, and transparency are crucial to ensuring that everyone understands the expectations, the process, and their role in it. Finally, flexibility is key. The timelines and protocols of academia differ from those of communities, making it essential to adapt and navigate these differences to co-produce outcomes that are meaningful and valuable to both parties.

8 Conclusion

> [Regarding transformations] As long as there are community groups and people who decide to work to transform the reality of a community, it is possible. Is it easy? No. But it is possible.[25]

Amid the escalating climate and ecological crisis, transformations toward sustainability are needed and require alternative socio-ecological interactions. This need is not for one-size-fits-all solutions, but rather for alternatives that consider both local and global challenges and that aim to generate meaningful and long-lasting change. *Casa Pueblo*'s experience is an example of how citizen-led initiatives articulate change through everyday actions. By exploring the "how" of their *autogestión*-driven transformation and focusing on community members' experiences, it is noticeable that community *autogestión* is not only a response to ongoing local challenges (e.g., climate change; disasters), but also an everyday practice of "deconstruction" of harmful ideas. In this path, diverse initiatives are developed (e.g., art; solar energy; forest visits) forming a "transformative whole" that challenges dominant narratives of colonialism and coloniality while working toward new possibilities. These initiatives are rooted in the colonial context of Puerto Rico, characterized by extraction, exploitation, and dependency, as well as the current context of vulnerability. As a result, they open the path toward alternative futures from the one imposed from above by redefining the relationship with nature and the idea of community and by encouraging the agency of everyone. In doing so, they demonstrate that citizen-led efforts can generate meaningful outcomes from the bottom up that can inform broad discussions on transformations.

Incorporating PV to explore *Casa Pueblo*'s community *autogestión* served as a means to advocate for community-engaged, context-sensitive, co-created methodologies that embrace multiple ways of knowing and acting. Although there are challenges related to participant engagement, including capacity/time and the need for economic and technical resources, it is worthwhile to incorporate these approaches to create spaces for diverse voices, views, and collaborations. In this context, as highlighted in the chapter, there is no exact recipe. Still, the goal is

[25] On-site film session, Participant 1.

to develop different ways of engaging with communities, conducting research, and presenting findings. Hence, using PV to film community *autogestión* served as a method of research and a collective practice of critical reflection and engagement, serving as a decolonial and future-oriented practice. This chapter shares insights on how filmmaking can facilitate the co-production of knowledge, shedding light on how academic and community expertise can collaborate, placing the researcher as facilitator rather than only expert. In doing so, it pushes toward a transformation in the ways of conducting research, aiming for creative and collaborative work that benefits both community and academia in the face of current social and environmental crises. As a participant stated in the quote above, it is not easy, but it is possible.

Acknowledgements López-González sincerely thanks *Casa Pueblo*, along with the participants of the project: Laura Irizarry, Jenniffer Pérez, Claudia Irizarry, Verónica Aponte, and Natalia Irizarry. Special thanks to Hery Colón and *Casa Comunitaria de Medios* in Aguirre at Salinas, Puerto Rico, for the technical support. The author is also grateful to Dr. Noella Gray for her advice and support, and to Dr. Mimi Sheller and Dr. Sarah Molinari for their valuable feedback.

Competing Interests This work draws on doctoral research supported by the Social Sciences and Humanities Research Council of Canada (SSHRC) (Grant Number – 430736) and the University of Guelph.

Ethics Approval This study was performed in line with the standards of the University of Guelph Research Ethics Board (REB 22-08-012).

References

Avilés-Vázquez, K. R., García-López, G. A., Gerena, C. E. R., Ortiz, E. M., Uriarte-Centeno, E. V., Ramírez, R. T., Vázquez-Negrón, J. J., Díaz, M. R., Valderrama, J. S., & Díaz, A. M. R. (2023). *Environmental Justice Movements as Movements for Life and Decolonization: Experiences from Puerto Rico*. Routledge.

Bai, X., van der Leeuw, S., O'Brien, K., Berkhout, F., Biermann, F., Brondizio, E. S., Cudennec, C., Dearing, J., Duraiappah, A., Glaser, M., Revkin, A., Steffen, W., & Syvitski, J. (2016). Plausible and Desirable Futures in the Anthropocene: A New Research Agenda. *Global Environmental Change, 39*, 351–362. https://doi.org/10.1016/j.gloenvcha.2015.09.017

Baldacchino, G. (2008). Studying Islands: On Whose Terms? Some Epistemological and Methodological Challenges to the Pursuit of Island Studies. *Island Studies Journal, 3*(1), 37–56.

Bennett, E. M., Solan, M., Biggs, R., McPhearson, T., Norström, A. V., Olsson, P., Pereira, L., Peterson, G. D., Raudsepp-Hearne, C., Biermann, F., Carpenter, S. R., Ellis, E. C., Hichert, T., Galaz, V., Lahsen, M., Milkoreit, M., Martin López, B., Nicholas, K. A., Preiser, R., & Xu, J. (2016). Bright Spots: Seeds of a Good Anthropocene. *Frontiers in Ecology and the Environment*, *14*(8), 441–448. https://doi.org/10.1002/fee.1309

Bentz, J., O'Brien, K., & Scoville-Simonds, M. (2022). Beyond "blah blah blah": Exploring the "How" of Transformation. *Sustainability Science, 17*(2), 497–506. https://doi.org/10.1007/s11625-022-01123-0

Bluwstein, J. (2021). Transformation is not a Metaphor. *Political Geography, 90*, Article 102450. https://doi.org/10.1016/j.polgeo.2021.102450

Cardinal, A. (2019). Participatory Video: An Apparatus for Ethically Researching Literacy, Power and Embodiment. *Computers and Composition, 53*, 34–46. https://doi.org/10.1016/j.compcom.2019.05.003

Cork, S., Alexandra, C., Alvarez-Romero, J. G., Bennett, E. M., Berbés-Blázquez, M., Bohensky, E., Bok, B., Costanza, R., Hashimoto, S., Hill, R., Inayatullah, S., Kok, K., Kuiper, J. J., Moglia, M., Pereira, L., Peterson, G., Weeks, R., & Wyborn, C. (2023). Exploring Alternative Futures in the Anthropocene. *Annual Review of Environment and Resources*, *48*, 25–54. https://doi.org/10.1146/annurev-environ-112321-095011

Crocker, S. (2003). The Fogo Process: Participatory Communication in a Globalizing World. *Participatory Video: Images that Transform and Empower*, 122–145.

Cui, M., & Brombal, D. (2023). From Resistance to Transformation – The Journey to Develop a Framework to Explore the Transformative Potential of Environmental Resistance Practices. *Philosophy & Social Criticism, 49*(5), 599–620. https://doi.org/10.1177/01914537231164186

Dunlap, A., & Tornel, C. (2023). Pluralizing Energy Justice? Towards Cultivating an Unruly, Autonomous and Insurrectionary Research Agenda. *Energy Research & Social Science, 103*, Article 103217.

Escobar, A. (2018). *Designs for the Pluriverse: Radical Interdependence, Autonomy, and the Making of Worlds*. Duke University Press.

Fisher, J. C., Mistry, J., Pierre, M. A., Yang, H., Harris, A., Hunte, N., Fernandes, D., Bicknell, J. E., & Davies, Z. G. (2021). Using Participatory Video to Share People's Experiences of Neotropical Urban Green and Blue Spaces with Decision-Makers. *The Geographical Journal*, *187*(4), 346–360.

García-López, G. A., Velicu, I., & D'Alisa, G. (2017). Performing Counter-Hegemonic Common(s) Senses: Rearticulating Democracy, Community and

Forests in Puerto Rico. *Capitalism Nature Socialism, 28*(3), 88–107. https://doi.org/10.1080/10455752.2017.1321026
Garriga-López, A. (2019). Puerto Rico: The Future In Question. *Shima: The International Journal of Research into Island Cultures, 13*(2). https://doi.org/10.21463/shima.13.2.13
Hudson, J. P. (2010). Formulaciones teórico-conceptuales de la autogestión. *Revista Mexicana de Sociología.*
Kindon, S., Pain, R., & Kesby, M. (2007). *Participatory Action Research Approaches and Methods*. Routledge.
Leeuw, S. D., Cameron, E. S., & Greenwood, M. L. (2012). Participatory and Community-Based Research, Indigenous Geographies, and the Spaces of Friendship: A Critical Engagement. *The Canadian Geographer/le Géographe Canadien, 56*(2), 180–194.
Lloréns, H. (2021). *Making Livable Worlds: Afro-Puerto Rican Women Building Environmental Justice* (1st ed.). University of Washington Press.
López-González, B. (2025a). From *Cucubano* to *Cucubano*: Facilitating Decolonial Energy Justice Through the Praxis of *Autogestión*. *Human Geography, 18*(2), 257–263. https://doi.org/10.1177/19427786251322837
López-González, B. (2025b). Art for Re-Existence: Casa Pueblo's Decolonial Landscape Interventions Toward an Energy Insurrection. *Human Geography, 18*(2), 272–280. https://doi.org/10.1177/19427786251319763
Lunch, N., & Lunch, C. (2006). *Insights into Participatory Video: A Handbook for the Field*. InsightShare.
Marzi, S. (2023). Co-Producing Impact-in-Process with Participatory Audio-Visual Research. *Area, 55*(2), 295–302. https://doi.org/10.1111/area.12851
Massol-Deyá, A. (2022). *Democracia y energía: Desafiando la economía de los combustibles fósiles por un país propio*. Ediciones Callejón.
Massol-González, A. (2024) *Autogestión Comunitaria: Casa Pueblo construye Modelo Comunitario Alternativo al sistema colonial en Puerto Rico*. Casa Pueblo Editorial.
Massol-González, A. (2022). *Casa Pueblo: A Puerto Rican Model of Self-Governance*. Lever Press.
Méndez, N., & Vallota, A. (2006). Una perspectiva anarquista de la autogestión. *Revista Venezolana de Economía y Ciencias Sociales, 12*(1), 59–72.
Mitchell, C., Lange, N. D., & Moletsane, R. (2018). *Participatory Visual Methodologies: Social Change*. SAGE Publications Ltd. https://doi.org/10.4135/9781526416117
Morales, T., Muñiz, S., Montero Sánchez, D., & Moreno-Domínguez, J. M. (2021). Autonomía y sostenibilidad en prácticas de vídeo participativo con comunidades indígenas. El caso de La Marabunta Filmadora. *Andamios, 18*(47), 397–422.

Moulton, A. (2024). Modernity's Antillean ecologies: Dispossession, disasters, justice, and repair across the Caribbean archipelago. *Progress in Environmental Geography, 27539687241236193.* https://doi.org/10.1177/27539687241236193

O'Brien, K., Carmona, R., Gram-Hanssen, I., Hochachka, G., Sygna, L., & Rosenberg, M. (2023). Fractal Approaches to Scaling Transformations to Sustainability. *Ambio, 52*(9), 1448–1461.

O'Brien, K., & Sygna, L. (2013). Responding to Climate Change: The Three Spheres of Transformation. *Proceedings of Transformation in a Changing Climate, 16*, 23.

Ortiz Gómez, M. G. (2013). ¿Neoliberalismo autogestivo? La Cultura de Autogestión para el Desarrollo como herramienta analítica. *Contextualizaciones Latinoamericanas*, (9).

Pelenc, J., Wallenborn, G., Milanesi, J., Sébastien, L., Vastenaekels, J., Lajarthe, F., Ballet, J., Cervera-Marzal, M., Carimentrand, A., Merveille, N., & Frère, B. (2019). Alternative and Resistance Movements: The Two Faces of Sustainability Transformations? *Ecological Economics, 159*, 373–378. https://doi.org/10.1016/j.ecolecon.2019.01.013

Post, E. (2023). Proyectos de muerte and proyectos de vida: Indigenous counter-hegemonic praxis to sustainable development in the Sierra Norte de Puebla, Mexico. *The Journal of Peasant Studies*, *50*(5), 1945–1974. https://doi.org/10.1080/03066150.2022.2082286

Rivas, M. (2018). La respuesta al desastre y las organizaciones sin fines de lucro en Puerto Rico. *Revista de Administración Pública, 49*, 39–65.

Rodríguez-Soto, I. (2020). Mutual Aid and Survival as Resistance in Puerto Rico. *NACLA Report on the Americas, 52*(3), 303–308. https://doi.org/10.1080/10714839.2020.1809099

Rodríguez Tamayo, N. A. (2019). La autogestión como resistencia, dos ejemplos en América Latina. *Kavilando, 11*(1), 119–139.

Rodríguez, I., Walter, M., & Temper, L. (Eds.). (2024). *Just Transformations: Grassroots Struggles for Alternative Futures*. Pluto Press.

Roque, A., Wutich, A., Brewis, A., Beresford, M., García-Quijano, C., Lloréns, H., & Jepson, W. (2021). *Autogestión* and Water Sharing Networks in Puerto Rico after Hurricane María. *Water International, 46*(6), 938–955. https://doi.org/10.1080/02508060.2021.1960103

Sandoval Rivera, J. C. A. (2017). Investigación colaborativa y decolonización metodológica con cámaras de video. *Universitas-XXI, Revista de Ciencias Sociales y Humanas, 27*, 161–177.

Serrano-García, I. (2020). Resilience, Coloniality, and Sovereign Acts: The Role of Community Activism. *American Journal of Community Psychology, 65*(1–2), 3–12. https://doi.org/10.1002/ajcp.12415

Soto Vega, K. (2019). Puerto Rico weathers the storm: Autogestión as a coalitional counter-praxis of survival. *Feral Feminisms, 9*, 39–55.
Stock, R. (2023). Abolition solarities: Theorizing antiracist and anticapitalist solar energy insurrections. *Renewable and Sustainable Energy Transition, 4*, 100063. https://doi.org/10.1016/j.rset.2023.100063
Torres-Abreu, A., García-López, G. A., & Concepción, C. M. (2023). De la protesta a la propuesta: Articulations between Environmental Justice Movements and Community-Based Management of Protected Areas in Puerto Rico. *Centro Journal, 35*(1), 23–55.
Temper, L., Walter, M., Rodriguez, I., Kothari, A., & Turhan, E. (2018). A Perspective on Radical Transformations to Sustainability: Resistances, Movements and Alternatives. *Sustainability Science, 13*(3), 747–764. https://doi.org/10.1007/s11625-018-0543-8
Tormos, F. (2018). The Politics of Survival in Puerto Rico: The Balance of Forces in the Wake of Hurricane María. *Alternautas*, 5(2), Article 2. https://doi.org/10.31273/alternautas.v5i2.1083
Vieta, M. (2014). The Stream of Self-Determination and Autogestión: Prefiguring Alternative Economic Realities. *Ephemera: Theory and Politics in Organization*, *14*(4), 781–809.
Villarrubia-Mendoza, J., & Vélez-Vélez, R. (2020). Centros de Apoyo Mutuo: Reconfigurando la asistencia en tiempos de desastre: Mutual Support Centers: Reconfiguring assistance in times of disaster. *Centro Journal, 32*(3), 89–117.
Vogel, C., & Brien, K. (2022). Getting to the Heart of Transformation. *Sustainability Science, 17*(2), 653–659. https://doi.org/10.1007/s11625-021-01016-8
Whiting, R., Symon, G., Roby, H., & Chamakiotis, P. (2018). Who's Behind the Lens? A Reflexive Analysis of Roles in Participatory Video Research. *Organizational Research Methods, 21*(2), 316–340.

Index

S. Molinari and M. Sheller (eds.), *Building Equitable Climate Adaptation Partnerships in the US Caribbean*,
https://doi.org/10.1007/978-3-032-15727-0

FSC
www.fsc.org
MIX
Papier | Fördert
gute Waldnutzung
FSC® C083411

Kolibri 360 GmbH
Ferdinand-Jühlke-Straße 7
99095 Erfurt, Deutschland
produktsicherheit@kolibri360.de